5th International Symposium on the Science and Technology of Light Sources

York, United Kingdom
10 - 14 September 1989

Editors:

Robin Devonshire
John Meads
David O. Wharmby

ISBN: 978-1-5108-4109-3

Printed from e-media with permission by:

Curran Associates, Inc.
57 Morehouse Lane
Red Hook, NY 12571

Some format issues inherent in the e-media version may also appear in this print version.

Printed by Curran Associates, Inc. (2017)

For permission requests, please contact Foundation for the Advancement of the Science & Technology of Light Sources (FAST-LS) at the address below.

Foundation for the Advancement of the Science & Technology of Light Sources
FAST-LS
Belmayne House
99 Clarkehouse Road
Sheffield, United Kingdom
S10 2LN

www.fast-ls.org

Additional copies of this publication are available from:

Curran Associates, Inc.
57 Morehouse Lane
Red Hook, NY 12571 USA
Phone: 845-758-0400
Fax: 845-758-2633
Email: curran@proceedings.com
Web: www.proceedings.com

Fifth International Symposium on the Science & Technology of Light Sources

York - England

10-14 September 1989.

Invited Papers

and

Abstracts.

LS : 5

Fifth International Symposium on the Science & Technology of Light Sources (LS:5)

Proceedings of the
Fifth International Symposium on the
Science & Technology of Light Sources (LS:5)
York - England
10 - 14 September 1989

Invited Papers
and
Abstracts of Landmark
and Poster Papers

Edited by:
R. Devonshire and J. Meads
High Temperature Science Laboratories
Department of Chemistry
University of Sheffield
and
D.O. Wharmby
Thorn Lighting Limited

Published by:
University of Sheffield

Preface

This volume contains both the full texts of the Invited papers, and the abstracts of the accepted Contributed papers, that are to be presented at the Fifth International Symposium on the Science and Technology of Light Sources (LS:5) being held in York, England, from the 10th to the 14th of September, 1989. The York Symposium has continued the trend established by earlier Symposia in the series by showing a further significant growth in terms of the numbers of participants (over 200) and the number and range of the contributed papers (109) compared to its predecessor. The previous Symposia were held in Loughborough (1975; 90 delegates), Enschede (1979; 126), Toulouse (1983; 150), and, most recently, in Karlsruhe (1986; 198).

The steady growth of interest in this meeting has placed constraints on the format of the Symposium itself. Apart from the closing session, which has a special format, the seven main sessions contained in the scientific programme of LS:5 comprise a long poster session bracketed by one, or more, Invited and/or "Landmark" papers presented to the full Symposium. The Landmark papers included in the programme were selected from the Contributed papers by the Organising Committee for presentation to the full Symposium because they met one, or more, of several criteria, for example, they were regarded as being either novel, timely, or of general importance in their particular session. In addition special consideration was given to the papers of younger scientists.

Each paper in the volume has been assigned a number according to its position in the running order (papers given in a particular poster session are given in the alphabetic order of the first author's name) and a further letter designation according to which of the three possible categories it belongs, i.e., I (Invited), L (Landmark), or P (Poster).

The Editors would like to give special thanks to Mike Duncan and Mark Fields (Sheffield University) for their help in producing the volume and to the Printing Unit at Sheffield University for meeting the tight printing schedule.

Robin Devonshire
John Meads
David O. Wharmby

July 1989

Editors

LS:5 Organising Committee.

*	Dr. R. Devonshire (Chairman)	University of Sheffield, U.K.
	Dr. H. Hoerster	Philips Research, West Germany.
	Dr. A.G. Jack	Philips Lighting B.V., Holland.
*	Mr. J. Meads	University of Sheffield, U.K.
	Prof. H-P. Popp	University of Karlsruhe, West Germany.
*	Prof. J.F. Raffle (Treasurer)	University of Loughborough, U.K.
	Dr. A. Rahal[1]	Universite Paul Sabatier, Toulouse, France.
*	Mr. P. Stott	Osram (GEC) Ltd., U.K.
*	Dr. D.O. Wharmby (Secretary)	Thorn Lighting Ltd., U.K.

(* - Members of the local Organising Committee).

Corresponding members.

Prof. Dr. L. Bartha	Academy of Sciences, Hungary.
Dr. R.S. Bergman	GE Lighting Business Group, U.S.A.
Dr. S. Berman	Lawrence Berkeley Laboratory, U.S.A.
Prof. Z. Cai	Fudan University, China.
Dr. Y. Hirose	Toshiba Corp., Japan.
Dr. S. Kamiya	Matsushita Electronics Corp., Japan.
Prof. D. Karabourniotis	University of Crete, Greece.
Mr. J. Kossakowski	CIE, Poland.
Dr. J. Maya	GTE Lighting Products Group, U.S.A.
Dr. L. Pitchford	GTE Laboratories, U.S.A.
Dr. V. Steinbruck	VEB Narva, East Germany.
Dr. D. Vida	Tungsram AG, Hungary.
Dr. J.F. Waymouth	Consultant, U.S.A.

---

1. The Committee has learnt with great regret of the untimely death of Dr. Rahal on the 17th July 1989.

LS:5 Financial Guarantors.

The LS:5 Organising Committee gratefully acknowledge the important rôle of the following companies who have acted as financial guarantors.

GE Lighting

GTE Products Corporation

Osram-GEC Ltd.

Philips Lighting

Thorn Lighting Ltd.

LS:5 Sponsors and Exhibitors.

Sponsors:

Bentham Instruments Ltd.

Coherent (U.K.) Ltd.

Spectra Physics (U.K.) Ltd.

Exhibitors:

APL Engineered Materials Inc.

Coherent (U.K.) Ltd.

Edwards High Vacuum Ltd.

Flow Simulation Ltd.

Glen Spectra Ltd.

Monolight Instruments Ltd.

Photon Control Ltd.

Plansee Metals Ltd.

SAES getters (G.B.) Ltd.

Westdeutsche Quarzscmelze GmbH.

Scientific Programme

Texts of Invited, Landmark and Poster Papers.

Session A

Chairman: J Maya

1:I Developing lamps for lighting systems . . . 1
M Koedam

Low Pressure Discharges

2:L Time-dependent modelling of the low-pressure Hg rare-gas discharge lamp taking into account the non-Maxwell distribution of electron energy . . . 12
Koichi Wani

3:P Rare-gas discharge fluorescent lamps suitable for industrial uses . . . 14
Yoshinori Anzai Takehiko Sakurai Takeo Saikatsu Takashi Osawa

4:P Compact fluorescent light sources on the base of non-traditional constructions and the principle of radiation excitation . . . 16
A S Fedorenko A A Prjitkov V F Dadonov V I Martjinov

5:P SPD predictions for blended fluorescent lamps . . . 18
L L Hope

6:P The 254nm radiation output in a Hg-Ar discharge between tubes in coaxial configuration. . . 20
E Kindel H Lange

7:P Radiation of Hg-Ar low pressure discharges . . . 22
J Kreher D Foerster W Meier

8:P Approximative method to describe the positive column of low-pressure discharges . . . 24
F Lantos

9:P Influence of noble gas atom excitation on low pressure Hg discharges . . . 26
G G Lister U Greb

10:P Nonequilibrium thermodynamical discussion of ionisation in breakdown and glow . . . 28
B Nyiri

11:P Deuterium lamp with improved life time . . . 30
W Schwarz P March

12:P Deuterium lamp for spectro-analytical devices with reduced noise level . . . 31
W Schwarz P March

13:P Calculations of fluorescent lamp parameters . . . 32
R Sonnenberg H Gundel

14:P Preliminary report on the novel technology of circular fluorescent lamp . . . 34
Mi Zhongliang

15:P On the experimental and theoretical determination of the electric field strength for the super-narrow tube diameter, low pressure Hg-Ar lamp . . . 35
G Zissis P Benetruy J J Damelincourt A Por

16:P Comments on reaction rate calculations in positive column of low pressure Hg-Ar discharges . . . 37
G Zissis A Alkaa M Yousfi J J Damelincourt

17:L Additional electrode heating during dimming of HF operated fluorescent lamps . . . 39
F A S Ligthart H ter Heyden L Kastelein

18:L Suspension flow-coating of fluorescent lamps: modelling of the drying step . . . 41
F W Altena M Westerhof A Boomkens M van Rijswick

19:I Review of compact fluorescent lamp R & D . . . 43
S Murayama

Scientific Programme

Session B: Chemistry and Materials

Chairman: A G Jack

20:I Chemical and physical modelling of metal halide discharges - today and tomorrow . . . 53
E Fischer E Schnedler

21:P Experimental investigations of Na reactions in high pressure sodium (HPS) lamps . . . 63
R K Datta

22:P Thermodynamic considerations on transport properties of halogen-filled flood-light lamps . . . 65
G Dittmer U Niemann

23:P Fluid flow modelling of a modified transpiration experiment designed to investigate electrode corrosion . . . 67
M Duncan J Meads R Devonshire N W O'Brien S A Mucklejohn

24:P A technique for the study of interactions of metal halides with gaseous impurities . . . 69
M Fields S A Mucklejohn N W O'Brien

25:P Laser-Raman spectroscopy of metal halide vapours at high temperature . . . 71
M Fields R Devonshire H G M Edwards V Fawcett

26:P Thermochemical data of scandium halides evaluated from vapour pressure measurements and reactions with quartz . . . 73
E Fischer R Kaiser U Niemann

27:P Chemical vapour transport and complexation in the NaI-ScI3 systems . . . 75
K Hilpert M Miller M Sai-Baba

28:P Immunization of quartz surface against crystallization and diffusion . . . 77
S Jonas A Kwatera W Ptak E Walasek B Lesiuk W Stanioch

29:P Measurements of sodium loss from fused silica metal halide arc tubes . . . 78
W M Keeffe V D Meyer

30:P Vapour pressure and thermochemical parameters of metal halides: InCl and TlBr . . . 80
S A Mucklejohn N W O'Brien T R Brumleve

31:P Impurities in high pressure mercury discharge lamps . . . 82
H-P Popp R Ascher

32:P Chemical analyses of sodium, mercury, aluminium and trace elements in high pressure sodium lamps . . . 84
G D A Solleveld

33:P New aspects in the chemistry of sodium and potassium high pressure lamps . . . N/A
D Veza J Rukavina D Fijan G Pichler

34:L High temperature seal operation in a multicomponent high pressure metal vapor lamp . . . 86
R Snellgrove E Wyner

35:I Chemical and materials problems affecting lamps . . . 88
D B Shinn

Session C: Electronics and Novel Devices

Chairman: J F Waymouth

36:I Review of electrodeless discharges for lighting . . . 98
D O Wharmby

37:P Power supply for 2 kW halogen lamp with continuous power control and sinusoidal input current . . . 108
M Albach

38:P Impedance and emission properties of capacitatively coupled Hg-Ar low pressure discharges . . . 110
C Beneking

39:P Rg_2* and RgCl* excimer emission from dielectric barrier discharges . . . 112
B Gellert B Eliasson U Kogelschatz

40:P A combined pulse controlled discharge with a positive V/I characteristic . . . 114
V A Godyak

41:P Radiation characteristic of flash- and arc-lamps for laser pumping . . . 116
P Greve

42:P Pulsed operation of high pressure sodium lamps . . . 118
K Guenther H G Kloss T Lehmann R Radtke F Serick

43:P Operation of low wattage electrodeless HID lamps . . . 120
D Hollister M Neiger S Berman

44:P High power microwave lamp for pumping a Nd-YAG laser . . . 123
G Lohmann H-P Popp M Neiger

45:P Novel circuits for high frequency operation of HID lamps . . . 125
E J P Mascarenhas A L G Edwards

46:P Efficient production of incoherent excimer radiation in dielectric barrier discharges . . . 127
H Muller M Neiger V Schorpp K Stockwald

47:P New high power arc lamps operating in the middle frequency range from 10-100kHz . . . 129
T Reichert K-J Dietz A Dohn U Kronert G Mayer M Neiger H-P Popp

48:P Circuitous model of gas discharge lamp . . . 131
Lian-cai Wang Li-xi Yang Ke-jing Qin Geng-xin Jia

49:P Electronic ballast for high pressure sodium gas discharge lamp . . . 132
A Wegener T Husgen

50:P Simplification of electronic ballast for self-ballasted fluorescent lamps . . . 134
N Yoshikawa T Imai

51:L UV radiation from Hg resonance lines by means of dielectric barrier discharges . . . 136
B Gellert B Eliasson U Kogelschatz

52:L Electromagnetic compatibility of lighting systems . . . 138
J R Bergervoet

53:I Progress in electronics for lamps and its forecast . . . 140
E Bolte

Session D: Coatings and Incandescent Lamps

Chairman: S Berman

54:L An aluminium transport cycle observed by multiphoton LIF in operating halogen lamps . . . 150
J R Woodward

55:P FTIR diagnostics of tungsten-halogen lamps . . . 152
L Bigio

56:P The investigation of getter-gas interaction in gas-filled incandescent lamps. . . 154
E Biro J Adam S Reznik O Lojka

57:P Fluid flow modelling as an aid in the design of incandescent light sources . . . 156
D G Green R Devonshire S A Mucklejohn N W O'Brien

58:P Non contacting precision measurement of thin wires. . . 158
K Guenther H G Kloss W Luedge

59:P Interaction of heated tungsten with its radiation. . . 160
A S Ivantzev M I Obmajikina V I Konjashkina

60:P Temperature influence upon optical parameters of the infrared reflecting layers . . . 162
T Kanabus W Stanioch

61:P Electric bulb based radiation IR microfurnace for fast sintering and epitaxy . . . 164
A Szymanski

62:P Production technology, some electrical and light parameters of incandescent lamps with filament of composite material on carbon base. . . 165
N S Vdovin M I Rogajilin A V Mordjuk

63:P Molybdenum blackening in glass halogen lamps. . . 167
T H Yu J M Ranish

64:P Studies about halogen regenerate cycle agent in halogen lamp . . . 169
Wei-Zhong Zhou Shan-Chun Cao

65:L Compact quartz halogen lamps with infrared reflecting coatings . . . 171
R S Bergman

66:I Review of optical coatings for incandescent and other lamps . . . 173
Y Yuge

Session E: Electrodes and Diagnostics

Chairwoman: L C Pitchford

67:L Dynamics of the neutral and ionized barium atoms in the vicinity of fluorescent lamp electrodes . . . 181
A K Bhattacharya A S Awadallah

68:P The vaporization behaviour of pure BaO and alkaline earth containing mixed oxides . . . 183
L Bencze A Popovic O Kaposi

69:P Simulation of the fluorescent lamp cathode . . . 185
A K Bhattacharya J H Ingold T F Soules

70:P Thallium pressure determination from the Tl-377.6 nm resonance line in high pressure mercury + thallium iodide discharges . . . 187
S Couris E Drakakis M Aubes

71:P Electrode temperature measurements in gas discharge lamps . . . 189
E H Curtis A G van der Kooi

72:P Evaporation rates from an oxide coated tungsten electrode in low pressure discharge devices . . . 191
R D Groves A K Bhattacharya

73:P Barium diagnostics in front of heated fluorescent lamp electrodes . . . 193
R Hilbig U Chittka

74:P Space and time dependent modelling of glow and thermionic cathode spots . . . 195
S Hollo B Nyiri

75:P Time modulation of sodium density in Na-HgI AC arc . . . 197
D Karabourniotis A Palladas A Tsakonas

76:P Function and surface properties of emissive coatings and their influence on high pressure gas-discharge lamps . . . 199
J M Korochkov A M Kokinov

77:P Estimation of convective arc bending by equivalent field method . . . 201
F Markus B Nyiri K Gambar

78:P Arc temperature measurement of high-pressure discharges in scattering enclosures . . . 203
S A Mucklejohn A T Rowley

79:P Zero field thermionic emission in HF operated fluorescent lamps . . . 205
H J W Schenkelaars J J Heuvelmans A van der Kooi J Schlejen

80:P Experimental diagnostics and physical conditions in discharge plasma of metal-halogen lamps . . . 207
V G Vdovin V A Vdovina A A Pustoshkin

81:P Spectra and characteristics of high pressure sodium-cadmium lamp . . . 209
D Veza J Rukavina D Azinovic G Pichler

82:L Electrode diagnostics using plasma-enhanced photoemission . . . 211
D R Woodward M B Schulman

83:I Electrode phenomena . . . 213
F Nagel

Session F: High Pressure Discharges

Chairman: E Schnedler

84:L Stabilisation of white light SON lamps . . . 221
R A J Keijser

85:L Instant light metal halide lamp . . . 223
R H Springer R L Hansler J M Davenport

86:P A study of three dimensional natural convection in horizontal high pressure mercury lamps . . . 224
P Y Chang W Shyy J T Dakin M E Duffy

87:P A computer model of a high pressure discharge lamp based on the FLUENT code . . . 225
L Cifuentes D A J Mottram S Vasquez

88:P Computer predictions of flow patterns and temperature contours in high pressure discharge lamps . . . 227
L Cifuentes

89:P Three dimensional computer predictions of a high pressure discharge lamp. . . 229
L Cifuentes D A J Mottram

90:P Influence of the two-temperature hypothesis on the electrical characteristics of high pressure mercury discharge plasma . . . 231
S Couris G Zissis J J Damelincourt

91:P Determination of the total escape factor in high pressure mercury lamps . . . 233
J J Damelincourt E Drakakis J Salon E Roaux

92:P A long life metal halide lamp with variable electrical power . . . 235
K -J Dietz A Schonlein

93:P A Hg high pressure lamp with improved UV-C output . . . 236
K -J Dietz A Schonlein W Schwarz

94:P Ionic demixing in metal halide lamps . . . 237
W M Keeffe R W Liebermann H L Rothwell Jr

95:P A simplified radial temperature profile model for wall stabilized arc discharges and its application to the modelling of radiative spectra . . . 239
R W Liebermann

96:P Development and application of a 1-D model of a metal halide discharge lamp including detailed spectral predictions. . . 241
D A J Mottram L Cifuentes S A Mucklejohn

97:P A new metal halide lamp for color LCD projectors . . . 243
M Narita M Ohyama T Hiramoto

98:P Numerical simulation of rare gas high pressure short arc lamps . . . 244
A C Palacin M Neiger

99:P The future of metal-halide lamps for photochemical processes . . . 246
G A Sarytchev L B Prikupets G N Gavrilkina L A Smorodinova

100:P Run-up characteristics and photometric aspects of a metal-halide lamp for a motorcar headlamp system . . . 248
D van Pijkeren

101:P Improvement of ultraviolet-related properties of MHN-T lamps . . . 250
J A J M van Vliet W Verbeek

102:P Numerical tests of a simple empirical model for radiation transport in optically thick plasmas . . . 252
P A Vicharelli D Karabourniotis

103:L Computational fluid dynamics of high intensity arcs . . . 254
P A Vicharelli T Fohl

104:I Predicting high-pressure discharge performance. . . 256
J T Dakin

Session G: Closing Session

Chairman: R Devonshire

105:L Mercury consumption in fluorescent lamps . . . 266
P W C Verhees

106:L Computerized tomography of high intensity arcs . . . 268
T Fohl P A Vicharelli

107:I Where will the next generation of lamps come from? . . . 270
J F Waymouth

Post-Deadline Papers

108:P Characteristics of powers supplies for standard lamps . . . 276

E M Wareham

DEVELOPING LAMPS FOR LIGHTING SYSTEMS

M. Koedam
Philips Lighting B.V.
Central Development Laboratories
Eindhoven
The Netherlands

1. Introduction

The history of electric lighting started with the invention of the incandescent lamp more than 100 years ago.

Very many installations for indoor and outdoor lighting as well as for special purposes have been built since Edisons first demonstration. In the design of these installations the lamp was an invariable factor. Reflectors and lenses could be added as aids to bring the light at the right place.

In the course of time many new lamp types have been added to the original general purpose incandescent lamp. Today's lighting engineer has a wide choice and every year new lamp types are added while seldom a lamp type disappears. The basis of a lighting installation is a unit consisting of a burner, a control gear and optical aids such as a reflector and/or a lens.

Too often the development of such a unit starts with the burner. When the burner development has been completed the control gear has to be designed followed by the development of the optical system and the luminaire. The importance of the luminaire, is often underestimated. It is certainly more than an attractive looking metal case. Its design determines the operational tempeature of all components which it houses and thus the overall performance of the system.

A sequencial development as described above never results in the optimum solution. However, when only magnetic ballasts are used there is little flexibility in the system of burner and ballast. The same holds more or less for the optical system if the design work has to be done on a drawing board, which limits the complexity of the design.

However, nowadays computer aided design (CAD) can be used to optimize the burner/reflector/lens system, while electronic control gear offer the possibility of a free choice of arc voltage, current shape and frequency.

The development process has to be changed to make full use of these new possibilities. Lamp development has to be subordinated to the development of the system.

As a consequence lamp development is no longer the playground of physicists and chemists : It has to be a well organized multidisciplinary activity.

In this presentation some examples will be given of the system approach. The following items will be discussed.

* Development of incandescent reflector lamps.
* Application of interference filters on lamps.
* Integration of burner, control gear and ignitor in compact fluorescent lamps.
* Design of a new motorcar headlamp system.
* Development of a new system for outdoor sports lighting.
* Development of systems for accent lighting.

2. Development of incandescent reflector lamps

In most reflector lamps parabolic mirrors are used, even though these reflectors only give good beam shapes with a point light source and incandescent filaments are far from point sources. Moreover the position of the filament with respect to the parabolic reflector has never been optimized in present day reflector lamps.

The light distribution of reflector lamps can be improved by a different positioning of the filament and a carefully designed reflector.

In a recent publication Gehring indicates that the maximum beam intensity I_{max}, can be improved by a new reflector design, a slight variation of the filament pitch and a better position of the filament in the reflector. In this way the light distribution in the beam is also improved and stray light is reduced.

The optimal reflector shape and position of the filament in the reflector are calculated using CAD techniques.

Some provisional results for experimental 120V lamps achieved in this way are given in table 2.1.

Table 2.1.

lamp	lamp	existing R-type		new R-type	
diameter	power	I_{max}	beam spread	I_{max}	beam spread
50 mm	40W	350cd	35°	500cd	30°
60	60W	-	-	1000cd	25°
63[5]	60W	800cd	30°	-	-
80	75W	800cd	35°	1500cd	25°
95	100W	1500cd	30°	3300cd	20°

Figure 2.2. shows the relatively complex lamp geometry, which is necessary to achieve the required performance. The reflector is a metal layer deposited on the bulb wall. If, for the same bulb shape, a reflector layer is deposited on the top, a bowl reflectorlamp is obtained. The beam characteristics which can be obtained with this new bowl reflector are much better than with the present spherical bowl reflector lamps.

Table 2.3. shows the improvements in different luminaires

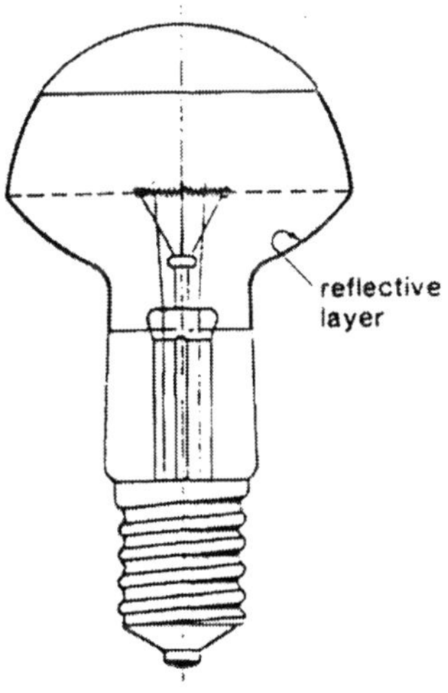

Fig. 2.2.

Table 2.3.

luminaire type	I_{max} present bowl reflector lamp	I_{max} new design bowl reflector lamp
1	2100 cd	3300 cd
2	2700 cd	4400 cd
3	3900 cd	6500 cd

CAD techniques made it possible to optimize the system of filament and reflector. The maximum beam intensity and the light distribution in the beam could be improved considerably.

3. Integration of halogen incandescent lamps and interference filters

Due to technological developments in the past years interference filters have become available for lamp coating. It proved to be possible to design filters that can be used up to temperatures of 800°C.

When these filters are designed to reflect infra-red radiation back to the filament and when the filament construction and positioning is optimised to absorb this radiation, then up to 40% improvement of the luminous efficacy can be obtained at constant lamp life. This result has been realized in commercially available 1500W lamps. The coated lamp gave the same luminous flux as the 1500W lamp at only 900W input power.

Interference filters can also be used to modify the spectrum. Examples of these filter application are :

- Filters that minimize the UV output of lamps.
- Coloured filters for instance to realize yellow motor car lamps, which are used in France.

Three techniques can be used for these lamp coatings:

- evaporation
- chemical vapour deposition
- dipping

Some of these techniques open the possibility to deposite a uniform coating even on complex surfaces. This

considerably enlarges the application possibilities of integrated systems based on filament/filter lamps.
Many new developments in these fields can be expected in the coming years. Also in this case the system approach where filament construction, bulb shape, reflection/ transmission curve of the filter and position of the filament with respect to the bulb can be adjusted, offers better possibilities for optimization than a construction with separate components.

4. Integration of burner, ignitor and control gear in compact fluorescent lamps

Since the energy crisis compact fluorescent lamps have been introduced for incandescent lamp replacement. To create a real retrofit, the ballast, ignitor and burner have to be integrated to a system which, in the same way as an incandescent lamp, can be connected to the mains.

The first integrated compact fluorescent lamp had a magnetic ballast and a glow switch for ignition. Later electronic control gear, working at 20-50 kHz, has been used. At these frequencies the anode fall of the gas discharge disappears and, as a consequence, the system efficacy increases by 10%. The lifetime of some electronic components is very sensitive to temperature. To guarentee a long lifetime the circuit lay-out and the position of the circuit with respect to the burner has to be carefully designed to achieve the lowest possible temperatures for the critical components. The heat flows in the system are of great importance since they determine the temperatures in the system.

The advantages of these electronic compact fluorscent lamps are : 10% higher luminous efficacy and lower weight. Reduction of the dimensions is mostly not possible, due to the temperature sensitivity.

The next step in the development of compact fluorescent lamps for incandescent replacement has been demonstrated by Hollister in 1976, following ideas of Anderson described in a patent application of 1968. He frequency of 13.6 MHz to operate an electrodeless compact fluorescent lamp.

In the Philips Lighting Laboratories an electrodeless compact fluorescent lamp concept has also been worked out, using the system shown in fig. 4.1. The HF power at 2.65 MHz is very efficiently coupled into the discharge using a cylindrical ferrite core. The discharge vessel is a sphere of 6-7 cm diameter, the input power is about 20W and the system efficacy 65-70 lm/W. The electrodeless compact fluorescent lamp approximates the incandescent lamp in light distribution as well as in size.

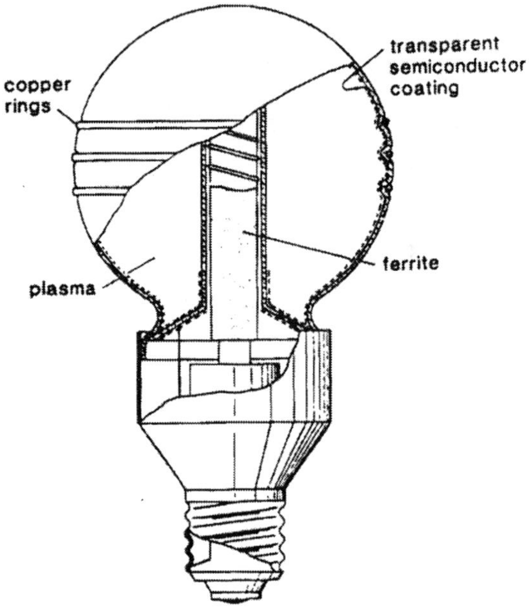

Fig. 4.1.

The development of this electrodeless compact fluorescent lamp is a clear example of system development. The electronic circuit and the burner have to be carefully tuned to each other to get maximum efficacy and minimum radio interference.

5. A new system for motorcar headlamp

Today halogen incandescent lamps are used in motorcar head lamp systems. The beam is formed using a parabolic reflector and a complicated lens. The performance of the system can be improved through improvements of either the optical system and/or the burner. The first way to improve the burner is by increasing the luminance of the filament. However, with a tungsten filament the possibilities are limited. An increase of the filament luminance decreases the lifetime.

The luminance of a HID arc is many times higher than the luminance of a tungsten filament. This higher luminance can be used in a motorcar headlamp :

(a) to improve the beam quality,

(b) to increase the intensities in the beam,

(c) to reduce the size of the lantern,

The automotive industry is interested in a reduction of the reflector height, which offers more freedom for the design of the car.

To obtain an optimal headlamp the reflector, lens and burner have to be developed as one system, since the beam specifications do not allow much latitude. The influence of a variation of one of the components has to be compensated to fulfill the beam regulations.

There have been a number of publications concerning the application of HID lamps in motorcar lanterns.

G. Szekacs suggested (J.I.E.S. 1986) a high pressure sodium lamp. However, in this case the luminance of the arc is reduced by the diffuse p.c.a. wall. Moreover, the colour temperature is too low and the chromaticity coordinates are not in the specified region. HPS lamps are not a good alternative for halogen incandescent lamps.

With metal halide discharges a spectral power distribution can be realized that fulfills the chromaticity requirements for motorcar headlamps. With quartz discharge tubes, the required high luminance can easily be achieved. The arc dimensions can be chosen to fulfill the requirements of the optical system. A first target was to obtain life times of the burner and of the other components of the head lamp system equal to the average life time of the car.

Lamp filling and burner size are for a great deal determined by the beam requirements. The best results can be obtained if a burner has no exhaust tube, since the remaining tip is a disturbing element in the optical system.

A motorcar head lamp must have a very short run-up time. About full light within 2 seconds, as can be obtained with incandescent lamps is impossible with conventional gas discharge/ballast systems. However, as shown in the poster presentations at this conference of v. Pijkeren and Davenport, a proper gas filling together with a special control gear give run-up times which are close to those of incandescent lamps.

The development of motor car headlamp systems is a good example how system development works.

6. A new system for sports lighting

Another example of system development is given by a recently designed system for stadium lighting.

The specification of the lighting system are determined by :

(a) The requirements for TV coverage concerning : illuminance level, colour temperature and colour rendering index. Glare for both players and spectators should be kept to a minimum.

(b) The requirements concerning straylight as formulated by the people living in the neighbourhood and traffic approaching the stadium.

(c) The requirements of the owner of the stadium concerning the costs of the installation. The lamp masts form a considerable part of the investments. The lamp mast diameter depends on the wind resistance of the luminaires on top of the mast. For a given light output the wind resistance of the luminaire must be as low as possible. However, the luminaire size also depends on the lamp power and the maximum tolerable temperatures. Only a system approach leads to the optimum solution.

(d) The lighting engineer wants maximum flexibility. This can most easily be obtained if a number of reflectors can be mounted in the same housing. In this way different beam shapes can be generated with the same burner in the same housing.

TV coverage asks for a colour temperature of 5500 K and a CRI above 90. This can be realized with a metal halide discharge in a specific salt mix. The optical designer asks for a length to diameter ratio of the arc below 3 to obtain well defined beams. The requirements given under c) strongly limit the design freedom of the lamp development engineer. In the example described in this section per lighting point 150 klm has to be generated with a luminous efficacy of 80 lm/W this means 1800W lamp power. The arc diameter of this type of discharge is 8 mm and consequently the electrode distance must be ≤ 25mm.

An outer bulb is a disturbing element in a high performance optical system, which leads to another limitation of the lamp concept.

Lamp life and maintenance must be as good as possible. A maintenance of 85% at 1500 hrs was the starting point.

Table 6.1. summarizes the above described influence of the various components on each other.

TABLE 6.1: A SYSTEM FOR SPORTS LIGHTING

	gas discharge burner requirements	optical system requirements
T.V. coverage	T_c= 5500 K C.R.I. > 90	E_{max}/E_{min} = 0.73 E_{av} = 1500 lux
optical designer	arc geometry: l/d < 3 applicable in a number of reflectors	
stadium owner	maintenance > 85% at 1500 hrs	minimum wind resistance of the luminaire taking into account maximum tolerable temperatures
environment		minimum stray light outside the stadium

Lamp development in such a system approach differs considerably from the way followed up to now, where the lamp was the starting point and the optical system has to be adapted to that lamp.

7. New systems for accent lighting

For accent and down lighting it is necessary to create in the most efficient way a light beam of a given shape with light of the desired colour temperature and CRI. The shape of the reflector depends on the geometry of the burner.
Following a system approach the combination lamp and reflector must be optimised. CAD techniques can be of great help in the optimising process. In the case of a saturated gas discharge lamp also the extra heating of the burner through back reflection must be taken into account.

To maintain the light quality a control unit has to be added to the gear. This unit controls the lamp current in such a way that the colour temperature and/or luminous flux are kept constant. The application of such a control unit has already successfully been demonstrated for super high pressure sodium lamps. More details are given in the contribution by Keijser at this conference.

The system consisting of lamp, control gear reflector has to be developed as a complete unit.

8. Conclusions

In the past the development of the components of a lighting system (burner, control gear and optical system) took place separately, mostly starting with the lamp. Seldom one organization had the tools and the experience for the development of the total system. However, in my opinion the best solution can only be found if a system approach is followed.

The availability of CAD systems offers the possibility for a optimum design of the optical system in relation to the burner geometry and the beam requirements.

Developments in the field of electronic circuitry for lamp stabilization, ignition and control open the possibility for more flexibility in the optimization of the lamp with or without special features.

In the future the development of new light sources will be replaced by the development of new lighting systems tailored for a specified application area.

Time-dependent modeling of the low-pressure Hg-rare gas discharge taking into account the non-Maxwell distribution of electron energy

Koichi Wani
Electronics Research Laboratory, Matsushita Electronics Corporation
1-1 Saiwai-cho, Takatsuki, Osaka 569, Japan

1. Scope

For low-pressure mercury rare-gas discharges, the growing application of fluorescent lamps pushed forward recent improvements in the modeling technique of the discharge plasma[1)-3)]. The author developed a time-dependent model for the low-pressure positive column taking into account the deviation of electron energy distribution from the Maxwellian. The results are found to be closer to the experimental results comparing with those under the Maxwellian approximation. The model is applicable to the fluorescent lamp discharges operated on both dc power source and ac power source up to 50 kHz.

2. Model

For the operating circuit with a choke ballast L (Fig.1), the temporal behavior of the lamp characteristics is given with differential equations shown in Fig.2[4)5)]. A 6-state model having 4 excited states is employed for the mercury atom.

The electron energy distribution function (EEDF) were calculated from the two-term Legendre polynomial expasion of the Boltzmann equation[6)]. The results were expressed by a two-temperature distribution in which the bulk temperature T_b represents the low energy part of the EEDF and the tail temperature T_t, the high energy part. In Fig.3, the ratios $r_T=T_t/T_b$ obtained by fitting two-temperature distributions to the results of Boltzmann equation analysis are plotted for several sets of conditions. As shown with the solid lines in Fig.3, the ratio r_T is empirically expressed as a function of electron density times electric field strength (n_exE) for a given set of gas density. The time consuming process in solving the Boltzmann equation at each time step of the differential equations is now replaced by the following simple procedure:

(1) calculate the average electron energy U_e from Eq.(1),
(2) obtain the ratio r_T by refering to Fig.3, and
(3) calculate bulk and tail temperatures as $T_b=2U_e/3k$ and $T_t=r_T\ T_b$.

3. Results and discussion

Calculated field strength E for a dc operated positive column is shown in Fig 4. The present calculation agrees with the experimental results (circles) around the operating condition (Tw=42°C) of the fluorescent lamp. Examples of the calculated lamp voltage and current waveforms for ac operations are shown in Fig.5a for 60 Hz and Fig.5b for 50 kHz. The present model gives an accurate estimation of the lamp characteristics and is applicable to a computer aided design for the flourescent lamps.

References

1. L.Vriens, J.Appl.Phys. 44, 3980. (1973).
2. R.B.Winkler et al., Beitr.Plasmaphys. 22, 401 (1982).
3. R.Sonnenburg and H. Grundel, Contrib.Plasma Phys. 27, 37 (1987).
4. K.Wani, J.Appl.Phys. 58, 2968 (1985).
5. K.Wani, National Tech.Rep. 33, 90 (1987).
6. K.Wani, J.Appl.Phys. 63, 5683 (1988).

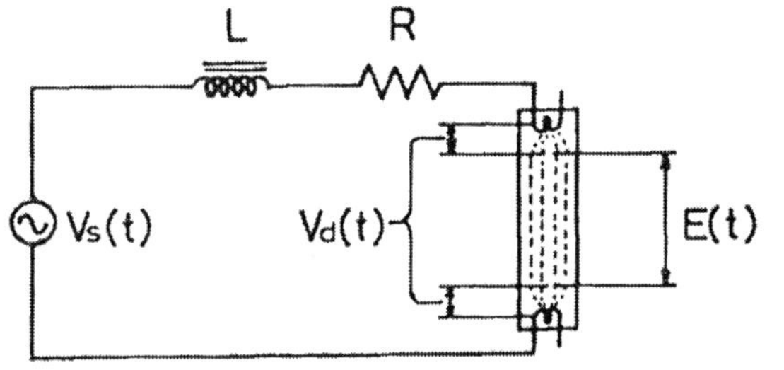

Fig.1 Lamp operating circuit.

$$\frac{d}{dt}U_e(t) = \frac{J^2}{n_e^2 c\mu_e} - \sum_{3}^{5}(-U_e + cV_i)S_i n_i - \nu_e \frac{2m_e}{m_g}(U_e - U_g) + \sum cV_{ij}C_{ij}n_i. \quad (1)$$

$$\frac{d}{dt}n_j(t) = \sum C_{ij}n_i n_e - \sum C_{ji}n_j n_e - S_j n_j n_e - A_{jg}^{*}n_j. \quad (2)$$

$$\frac{d}{dt}n_e(t) = \sum S_i n_i n_e + K_1 n_r n_r + K_2 n_p n_q - D_a \nabla^2 n_e. \quad (3)$$

$$L\frac{d}{dt}I(t) = -RI - V_d(t) - \frac{J\ell}{n_e^2 c\mu_e} + V_s(t). \quad (4)$$

Fig.2 Model equations.

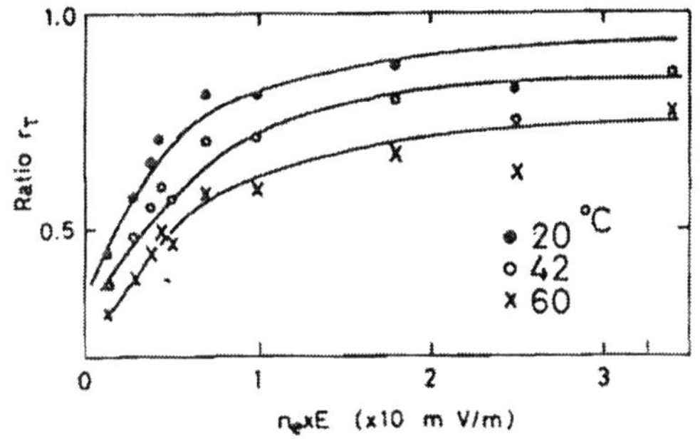

Fig.3 Ratio of T_t to T_b. Ar 400 Pa

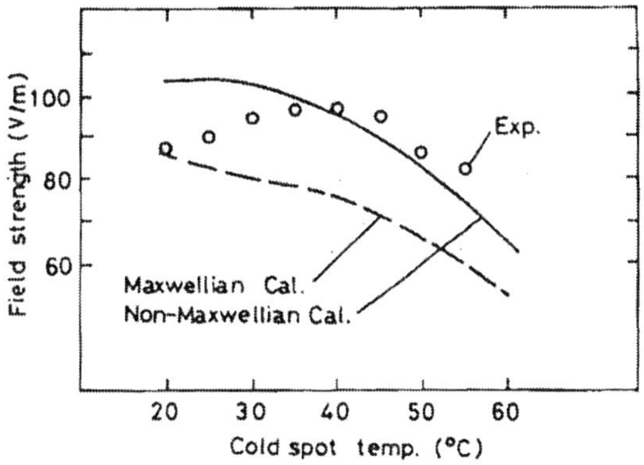

Fig.4 Electric field strength. d=24 mm

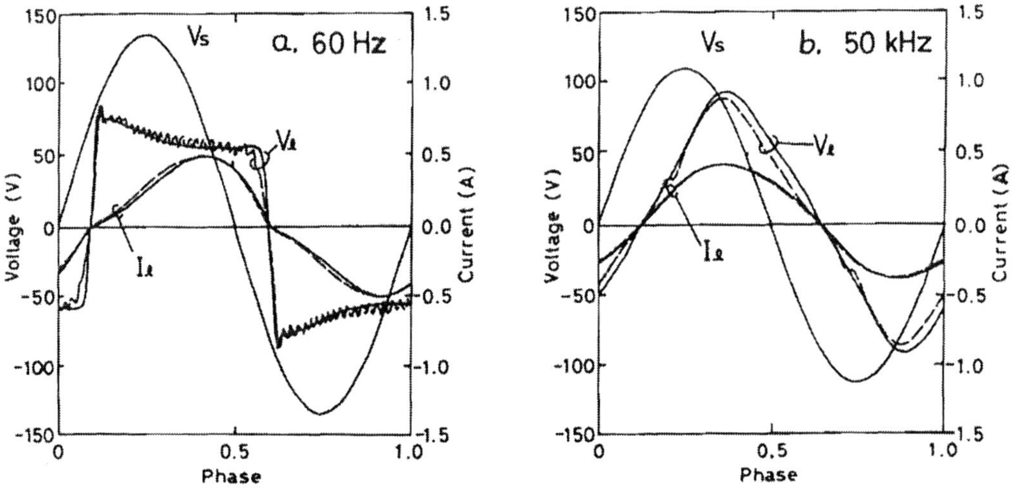

Fig.5 Lamp voltage and current waveform. d=24 mm, ℓ=500 mm.

Rare-Gas Discharge Fluorescent Lamps Suitable for Industrial Uses

Yoshinori Anzai, Takehiko Sakurai, Takeo Saikatsu and *Takashi Osawa
Mitsubishi Electric Corporation
Consumer Products Development Laboratory
2-14-40, Ofuna, Kamakura 247, Japan
* OSLAM-MELCO Ltd,
Ofuna Works, 5-1-1, Ofuna, Kamakura 247, Japan

Abstract

Fluorescent lamps have been widely used for many industrial applications, for example, light sources for duplicators, facsimiles, image reader etc. .

But, the fluorescent lamps have some inevitable defects ;e.g., temperature dependence and slow responce time of light output.

In order to improve the above mentioned problems, new rare-gas discharge fluorescent lamps have been investigated. These lamps have used rare-gases such as neon, argon, krypton and xenon as a substitute for mercury vapor in exciting phosphors within a discharge lamp.

The resonance radiation of the rare-gas discharge fluorescent lamp occurred in the far-ultraviolet, that is the well known Schuman region, which made for relatively low energy conversion from ultraviolet to visible light and hence low efficiency compared to the mercury discharge fluorescent lamps.

The lamp dimension, the gas filling condition and the lamp operating condition have been studied in order to obtain high light output and good performance.

Consequently we have discovered that the efficacy of the rare-gas discharge fluorescent lamp was increased by means of the operation of a pulse mode and the optimum selection of a gas filling pressure. The efficacy during the pulse was increased about two times as showh in Fig.1 and Fig.2.

These lamps showed instantaneous start-up time and largely independent of temperature-induced fluctuation. These new lamps have been applied to some practical uses and have given good results.

In near future, the rare-gas discharge fluorescent lamps will be applied to considerable industrial uses instead of a common fluorescent lamp.

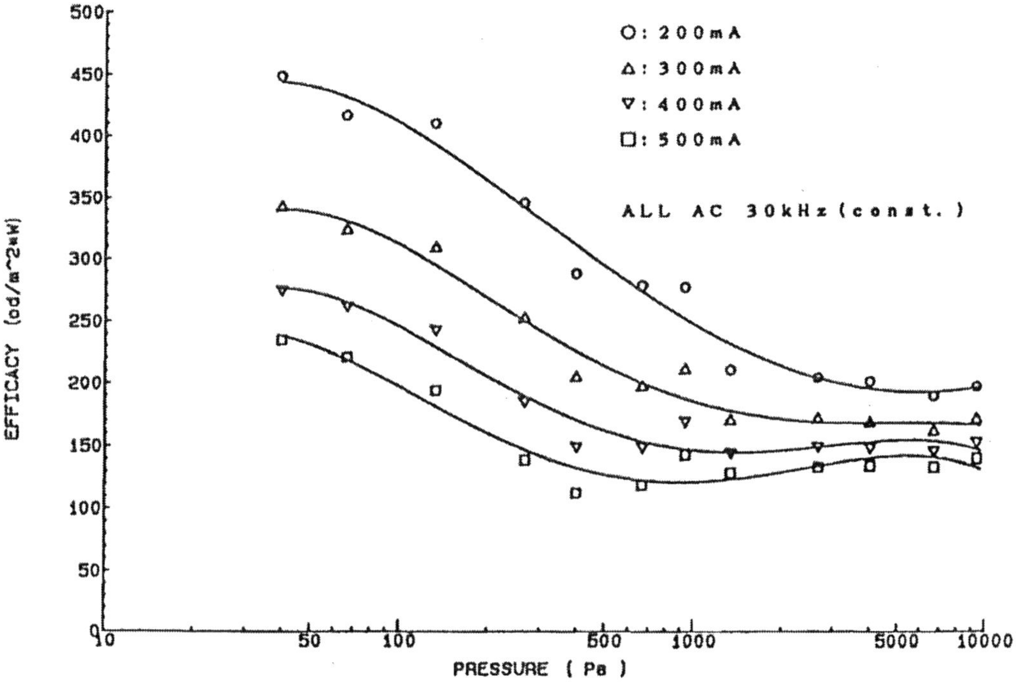

Fig. 1 The dependence of the efficacy on the filling gas pressure as a parameter of the lamp current. (Lamp diameter:15.5mm, Filling gas:Xe100%, Phosphor:$(Y,Sc)_2SiO_5$:Tb)

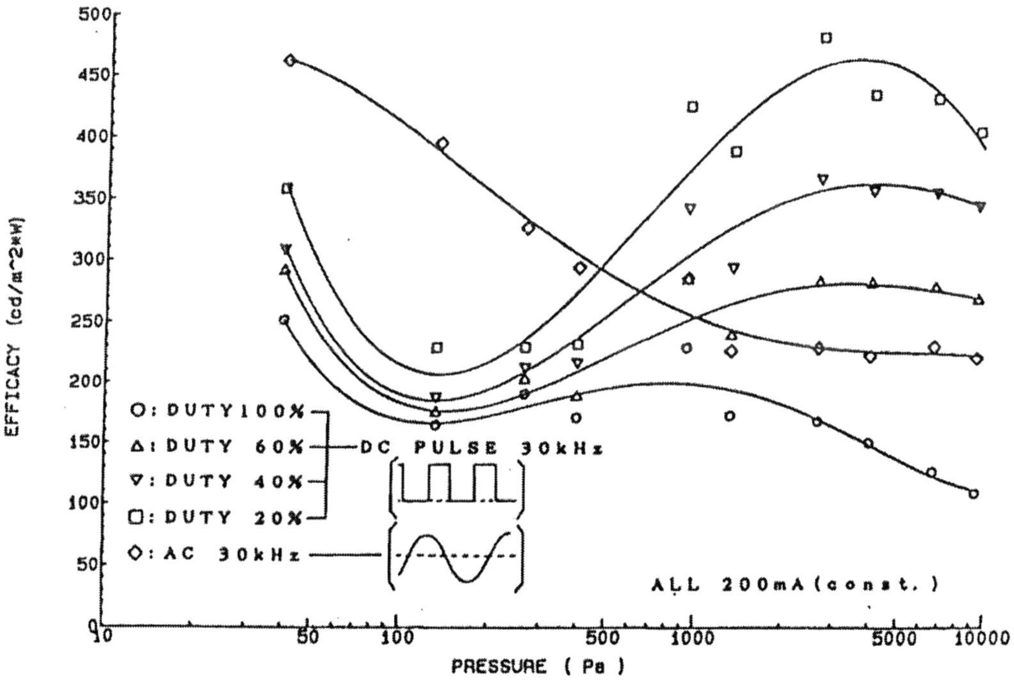

Fig. 2 The dependence of the efficacy on the filling gas pressure as a parameter of the duty ratio in the pulse mode. (Lamp diameter:15.5mm, Filling gas:Xe100%, Phosphor:$(Y,Sc)_2SiO_5$:Tb)

COMPACT FLUORESCENT LIGHT SOURCES ON THE BASE OF NON-TRADITIONAL CONSTRUCTIONS AND PRINCIPLE OF RADIATION EXCITATION

A.S.Fedorenko, A.A.Prjitkov, V.F.Dadonov, V.I.Martjinov, USSR

The usage of cathodoluminescence, high-frequency electromagnetic field for optical radiation production and also glassceramic elements opens new promising possibilities for non-traditional construction of light sources more effective then GLS incandescent lamps.

For cathodoluminescence light source with luminophor applied on inner glass tube surface coated by a film, anode, there stated the dependence of such characteristics of electron-beam discharge as firing voltage and anode current density on filling parameters. In this case we were succesful in decreasing firing voltage to 200 - 250V. Uniform anode distribution of electrons and tube surface brightness accordingly are achieved by cathode construction. Studied sample brightness was amounted to 200 - 1000cd/m^2, light output 25 - 30lm/W with power consumption of 3-2,5W.

For high-frequency electrodeless light source with h.f. oscillation generator, solenoidal inductor, sealed tube with inner cylinder that has an area for ionized medium and luminophor coating, it was stated that with argon fill,light output of generator-lamp can achieve 50lm/W and more,1500lm, excitation operating frequency -5-150mHz.

Electron concentration distributed in bulb is typical

ball-shaped with its maximum near the middle between inner and outer bulb cavity, symmetrically falling to bulb walls. Electron energy has marked minimum in the region of maximal concentration.

The use of glass-ceramic elements in compact fluorescent lamp construction enables to produce new highly effective lamp constructions including lamps with spot flux consisted of set of tubes coated by luminophor and lamps the sizes of which are proportionate with incandescent lamp sizes. Promising is also mechanical connection of linear tubes with all set of tubes placed in outer bulb, this procedure entirely eliminates tube welding or gluing together.

This type of lamps can operate from direct current circuit without cataphoresis manifestation.

SPD PREDICTION FOR BLENDED FLUORESCENT LAMPS

Lawrence L. Hope
GTE Products Corp
60 Boston Street
Salem, MA 01970 USA

A physical model that successfully estimates spectral power distributions for blended fluorescent lamps with arbitrary phosphor characteristics has been developed. It is known that the spectrum of the blended lamp is usually resolvable as a linear combination of the component spectra, i.e. the mercury visible spectrum and the spectra of the individual phosphors. The weighting factors to be assigned to the component spectra are not obvious and are derived here for the first time.

Common practice is use of the mass rule in which the SPD weighting factor for each component is presumed to be the fraction of the total phosphor mass due to that component. This is a good approximation only in special cases, i.e. blends in which absorption coefficients for ultraviolet are nearly the same for all phosphors used, and the general use of the mass rule has been the source of considerable confusion. It is replaced here by a dilution model that is supported by fundamental theory and has been verified experimentally. Properties of a single layer with homogeneous phosphor blend (good mixing) are treated, but the coefficients found may be used directly for calculation with lamps with multiple coatings.

The central derivation in the model is evaluation, for component i of the blend, of the effective scattering coefficient $\overline{\beta}_i$ and the effective absorption coefficient $\overline{\mu}_i$ from the properties and quantities present of the different phosphor types. The effective coefficients for the blended layer are sums of these. The uv absorption coefficient μ_i that component i would have if alone is diluted to a reduced value $\overline{\mu_i}$, with the dilution factor equal to the partial volume occupied by component i divided by the total phosphor volume. Contribution fractions to the lamp SPD are shown to be functions of the uv absorption coefficients and the mass densities in powder form.

The weight fractions obtained are approximately equal to the mass fraction only if the phosphors in the blend have similar uv absorption coefficients and mass densities. This is fortuitously true for many commercial phosphors, but is not true in general. <u>For widely differing uv absorption coefficients, use of weighting by mass to predict blend properties will produce serious errors.</u> The fraction of ultraviolet energy absorbed by component i will be $\overline{\mu}_i/\mu_B$. This is true even if the scattering coefficients for uv are very different for the separate components. Scattering will affect the total uv energy available at a given point, but whatever energy is available will be divided according to the diluted uv absorption coefficients only.

An experiment on a blend of two phosphors has provided excellent agreement with the new model in a case where the mass rule fails. Phosphor 1 is a green emitter with mass density in bulk $\rho_1 = 4.15$ g/cm^3 and uv absorption coefficient $\mu_1 = 0.30\ \mu m^{-1}$. Phosphor 2 is a red emitter with $\rho_2 = 5.00$ g/cm^3 and $\mu_2 = 0.19\ \mu m^{-1}$.

For two phosphors, a single parameter γ is defined that quantifies differential uv absorption. The simple mass weighting rule is obtained for the special case $\gamma = 1.0$. For the phosphors used, we obtain $\gamma = 0.527$. A series of lamps was prepared having percentages by mass of phosphor 1 (green) of 0, 20, 40, 60, 80, and 100%. Four lamps were made for each percentage. The chromaticity coordinates x and y for each lamp were measured. Plots of x and y as functions of the percentage of green phosphor were derived.

Figure 1 shows the x chromaticity coordinate vs percent mass of green phosphor for both models. The y chromaticity is similar but the observed points and the curve for $\gamma = 0.527$ lie above the mass rule curve. Agreement between experiment and the dilution model is quite good. The experimental points are far from the mass rule curve ($\gamma = 1$), and the mass rule must be rejected. Better agreement between observation and the mass rule is obtained for materials with closer values of ultraviolet absorption and mass density, but the agreement should be considered fortuitous. It is safer to use the dilution model for all cases.

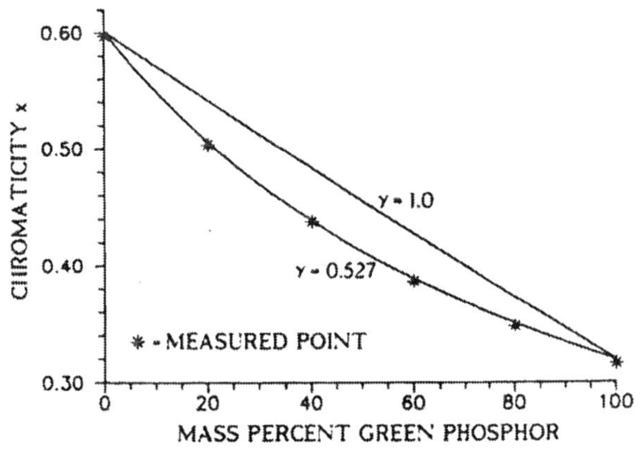

Figure 1. Chromaticity x vs Percent Green Phosphor
$\gamma = 1$ for Mass Rule and $\gamma = 0.527$ for Dilution Model

An estimate of the error for each model is the RMS deviation in x-y space. This is 0.09 for the mass rule, corresponding to a large, easily perceived color difference. Use of the dilution model reduces the error to 0.004, which is below perception threshold (inside the MacAdam ellipse).

A further experiment was carried out with blends of three phosphors. Two lamp series were fabricated, both containing the same blue component and the same red component, but not necessarily in the same amounts. The first series contained a standard green component, which was replaced in the second series by a another green phosphor. The two green phosphors provided similar performance when used alone in test lamps. They differ in uv absorption coefficient, which was 0.30 μm^{-1} for the standard material and 0.06 μm^{-1} for the replacement phosphor. The dilution model again allows much more accurate prediction. Average error is less than 1% for the dilution model but exceeds 25% for the mass rule.

The 254 nm-radiation output in a Hg-Ar-discharge between tubes in coaxial configuration

E. Kindel, H. Lange

Zentralinstitut fuer Elektronenphysik, Bereich Gasentladungsphysik, DDR - 2200 Greifswald, R.-Blum-Str. 8-10 in co-operation with VEB K NARVA, DDR, Berlin

The further development of fluorescent lamps to compact lamps makes positive columns with non circular cross sections of practical interest. One possibility for realizing those discharges is a bulb, which was made of an inner and outer tube in coaxial configuration. The electrodes sit at the ends of the two vertical sections, thus establishing a positive column throughout the volume between the inner and the outer tube.
We have calculated the radiation output, $P \sim \bar{n}/\tau_{eff}$, from the 254 nm-line of mercury of a dc discharge in Hg and 400 Pa Ar. Here is $\bar{n}$ the density of atoms in the 6^3P_1 level averaged over the discharge cross section and τ_{eff} is the effective radiative lifetime of the 6^3P_1 level, which both were measured with laser techniques, developed by /1,2/ for mercury discharges. τ_{eff} was determined by irradiating the discharge side on, with a dye laser pulse at 405 nm. This pulse generates a temporary overpopulation in the 6^3P_1 level. The following 254 nm fluorescence signal was recorded (Fig. 1a). For the density measurement of the 6^3P_1 level we used the hook method (Fig. 1b). In Fig. 2 τ_{eff} is plotted as a function of the wall temperature of the bulb, which determines the mercury vapour pressure. In this figure the results of the non circular cross section are compared with equivalent circular cross sections. Fig. 3 shows the value of n in the axis of the discharge. For the calculation of radially averaged densities we presumed a parabolic shape. The calculated radiation output is plotted in Fig. 4. The maximum of P for non circular cross section is shifted towards higher wall temperatures, because of the lower τ_{eff} compared to circular cross section. This can be caused by different radial distribution of excited atoms.

Reference

/1/ P.van de Weijer, R.M.M. Cremers J.Appl.Phys. 53,1401(1982)

/2/ P.van de Weijer, R.M.M. Cremers J.Appl.Phys. 57, 672(1985)

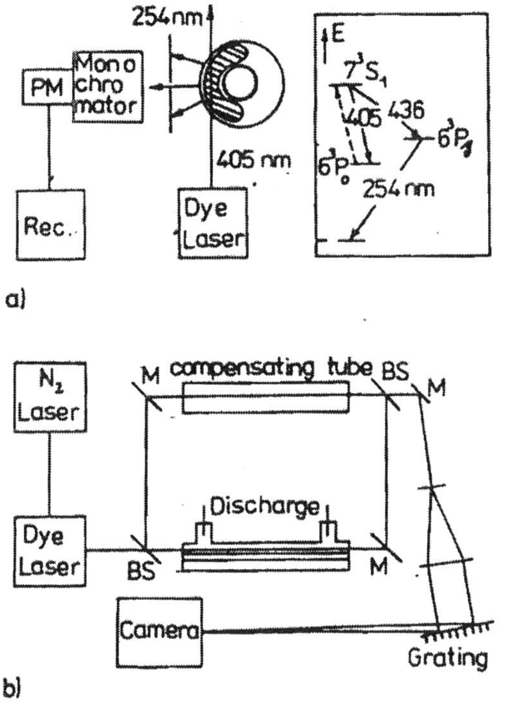

Fig.1 Experimental arrangment
a) τeff measurement
b) hook method

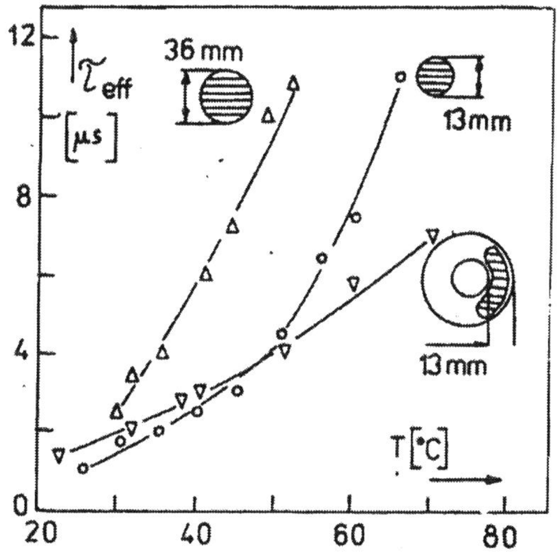

Fig.2 Effective lifetimes of the Hg 63P1-level in Hg + Ar

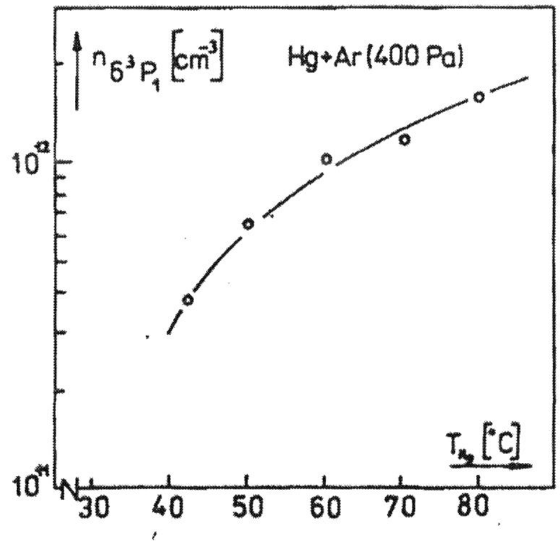

Fig.3 63P1-Density in the positive column with non circular cross section

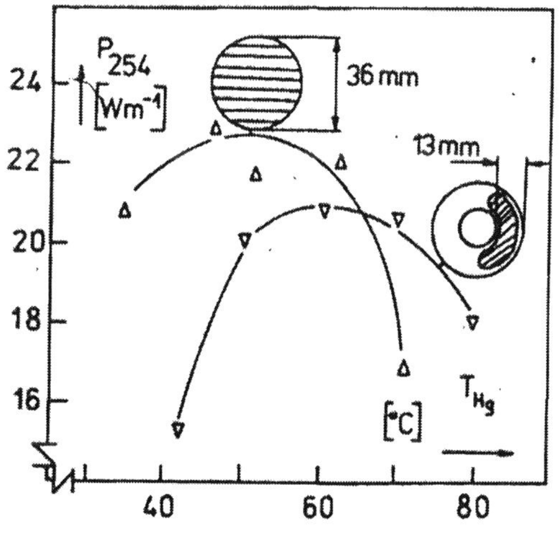

Fig.4 254 nm output as a function of the wall temperature

Radiation of Hg/Ar low pressure discharges

J. Kreher*, D. Foerster*, W. Meier**

* Zentralinstitut fuer Elektronenphysik, Bereich Gasentladungsphysik, DDR - 2200 Greifswald, R.-Blum-Str. 8-10

** VEB K NARVA, DDR - 1017 Berlin, Ehrenbergstr. 11-14

Fluorescent lamps are tending towards reduced tube diameters. The increasing wall losses lead to increased electric field strength E and affect the electron energy distribution function $f_e(U)$ and thus also collision processes and radiation efficiency η. Measurements of η for D = 10 mm and p_{Ar} = 800 Pa in dependence on the cold spot temperature T_{Hg} = 20 - 65°C and discharge current i = 32 - 400 mA were performed and the results (Figs. 1, 2) are compared with calculations of η (Fig. 4) using a system of balance equations for the excited Hg-atoms, for the discharge current and the electron Boltzmann equation. Former investigations /1/ compared ac-measurements with stationary calculations using effective E. Now measurements in dc-discharges allow a direct comparison. The maximum efficiency of the Hg-resonance line $\eta_{254\ nm}$ was > 50 % at T_{Hg} > 50°C, but $\eta_{185\ nm}$ monotonously increases with T_{Hg} in good accord with calculation: Fig. 4. The T_{Hg}-dependence of the visible Hg-lines (Fig. 2) show a monotonous increase with the Hg-density or a slight maximum (λ = 436, 365 nm). The Ar-lines show an revised behaviour e.g. 810, 840 nm: steep decrease with increasing T_{Hg} (Fig. 2). High E and low T_{Hg} = 20°C cause a predominant emission of Ar-lines: 1.1 W/m compared to 0.4 W/m of visible Hg-lines. The increase of the high energy part of measured $f_e(U)$ with decreasing T_{Hg} (Fig. 3) cause the increase of Ar-line emission (Fig. 2) and of inelastic collisions of electrons with Ar-atoms (Ar-inel) Fig. 4. Higher E increases the Hg-185 nm resonance radiation, increasing with T_{Hg} (experiment Fig. 1 and calculations Fig. 4). The higher emission of 185 nm and visible Hg and Ar-lines somewhat compensates the decrease of Hg-resonance line (254 nm) efficiency, and the maximum overall radiation efficiency reaches 75 %, which is sufficient for application in light sources.

/1/ J. Kreher et al. 19th ICPIG, Belgrad 1989, in print

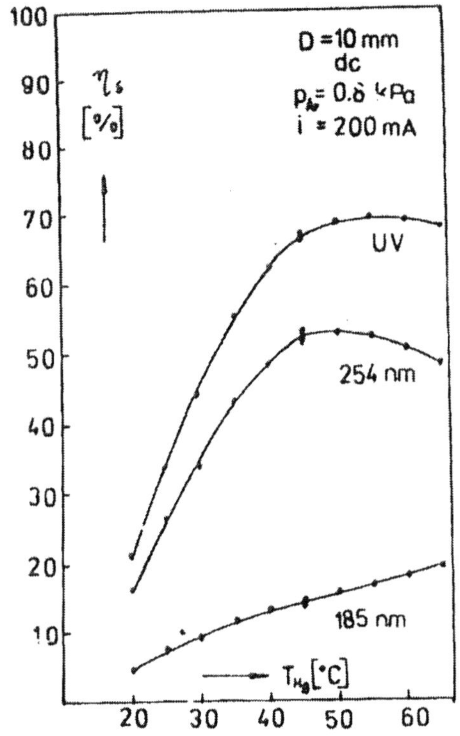

Fig.1: UV-efficiency

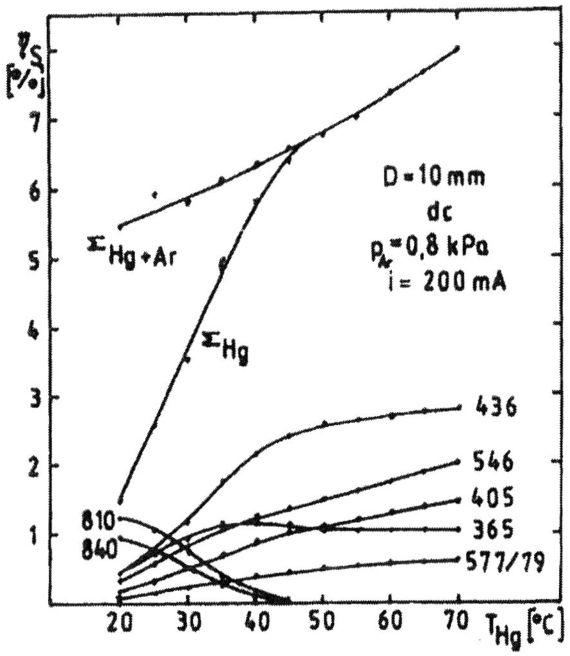

Fig.2: Visible and IR-radiation

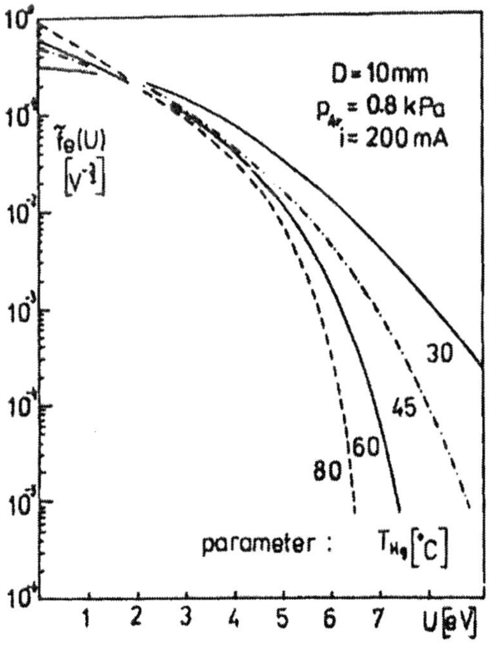

Fig.3: Measured $f_{\infty}(U)$

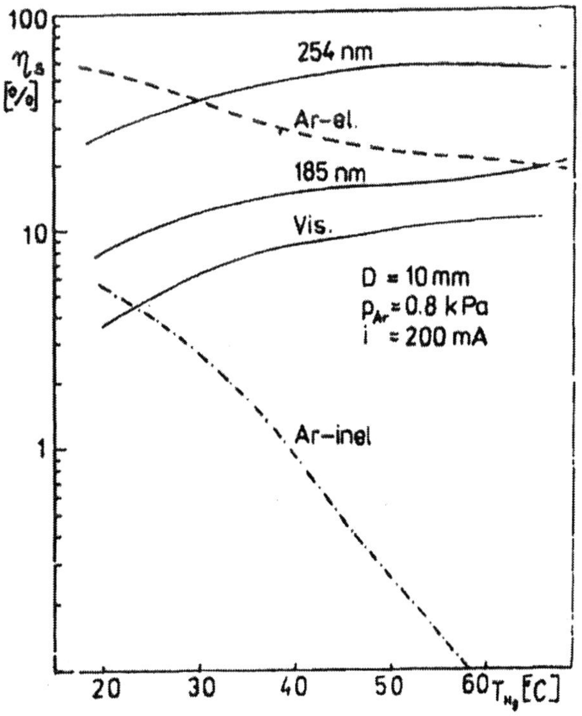

Fig.4: Results of calculations

Approximative Method to Describe the Positive Column of Low-Pressure Discharges

Ferenc Lantos Dipl. Phys.
TUNGSRAM RT, Váci út 77., 1340 Budapest, Hungary

Let us take the Boltzmann equation:

$$\partial f/\partial t + \underline{v}(\partial f/\partial \underline{r}) + \underline{F}/m(\partial f/\partial \underline{v}) = C(f, f_1, f_2, \ldots)$$

where f $\underline{r}, \underline{v}, t$ is the well-known Boltzmann function and $\underline{r}$, $\underline{v}$, t are the location, velocity and time in the same order. The C function describes the influence of the collisions between the particles. We can get equations to give the alteration of density of particles, impulse and energy at a certain point in the discharge by multiplying the original equation with 1, with $m\underline{v}$ or with $1/2\ mv^2$ and integrating over the velocity distribution of particles.

Let us presume that each types of particles is close to the local thermal equlibrium state /LTE/ at every points of positive column then:

$$f(\underline{r}, \underline{v}) = n(\underline{r})\left(m/2\pi kT\right)\ \exp\left(m(\underline{v}-\bar{\underline{v}})^2/2kT\right)$$

where n $\underline{r}$ is the particle density, T the temperature, m the mass of particle, k the Boltzmann constant and $\bar{\underline{v}}$ the average velocity. Furthermore, provided that:

$$m\bar{v}^2/2kT \ll 1$$

we can get the particle, impulse and energy density balance equations for the charged particles by integrating the Boltzmann equation over the particle velocity:

$$\mathrm{div}\ j = \eta$$
$$\pm en\underline{E} - \mathrm{grad}\ nkT = \Pi$$
$$5/2\ \mathrm{div}\,(jkT) - e\underline{E}j = \mathcal{E}$$

where j is the particle current density, $\underline{E}$ the electric field and $\eta, \Pi, \mathcal{E}$ are determined from the right side of the Boltzmann equation. They describe the influence of collisions.

In the case of low-pressure mercury-rare-gas discharge, the balance equations for the electrons:

$$\begin{aligned} &\mathrm{div}\ \underline{j}_e = \eta_e \\ &-en_e\underline{E} - \mathrm{grad}\,(n_e kT_e) = \underline{\Pi}_e \\ &(5/2)\,\mathrm{div}\,(\underline{j}_e kT_e) - e\underline{E}\underline{j}_e = \mathcal{E}_e \end{aligned}$$

For the Hg^+ or Hg_2^+ ions:

$$\begin{aligned} &\mathrm{div}\ \underline{j}_i = \eta_i \\ &en_i\underline{E} - \mathrm{grad}\,(n_i kT_i) = \underline{\Pi}_i \\ &(5/2)\,\mathrm{div}\,(\underline{j}_i kT_i) - e\underline{E}\underline{j}_i = 0 \end{aligned}$$

For the mercury atoms in the radiating, in the metastable or in the ground state:

$$\begin{aligned} &\mathrm{div}\ \underline{j}_m = \eta_m = 0 \\ &-\mathrm{grad}\,(n_m kT_m) = \underline{0} \\ &(5/2)\,\mathrm{div}\,(\underline{j}_m kT_m) = 0 \end{aligned}$$

For the rare gas:

$$\begin{aligned} &\mathrm{div}\ \underline{j}_r = 0 \\ &-\mathrm{grad}\,(n_r kT_r) = -\underline{\Pi}_e - \underline{\Pi}_i \\ &(5/2)\,\mathrm{div}\,(\underline{j}_r kT_r) = 0 \end{aligned}$$

With the aid of the described approximative method we can obtain expressions for the $_e$ and $_i$:

$$\begin{aligned} &\underline{\Pi}_e = \alpha_e \underline{j}_e = \left(A\sqrt{T_e} + B\,T_e + C\sqrt{T_e^3}\right) n_r \underline{j}_e \\ &\underline{\Pi}_i = \alpha_i \underline{j}_i = D\sqrt{T_i}\ \ n_r \underline{j}_i \end{aligned}$$

where A, B, C and D are constants determined from the cross section of collisions taking place between the electrons or ions and the rare gas atoms.

INFLUENCE OF NOBLE GAS ATOM EXCITATION ON LOW PRESSURE Hg DISCHARGES

G. G. Lister and U. Greb

Lamp Research, THORN Lighting Limited, Light Sources Division

Melton Road, Leicester LE4 7PD, U.K.

Numerical modelling of radial transport in low pressure discharge lamps was first used by Cayless, [1] to compute light efficiencies for discharges which contain a noble gas (or mixtures thereof) with a minority fraction of Hg vapour (typically $\sim 10^{-3}$). Since then, a number of other researchers have considered effects not included in [1], and the models have produced good agreement with experiment.

We have developed a computer code, based on the Cayless model, and extended to include effects considered important for the operation of low pressure discharge lamps at high powers. In particular, one of the effects of increased power input to these discharges is the reduction in Hg ground state atom density at the centre of the discharge. The processes causing this effect are gas heating and radial cataphoresis. Under such conditions, excitation of noble gas atoms are expected to be significant, but their effect has to date been ignored in the published literature.

Diffusion equations for five states of Hg, 6^1S_0, $6^3P_{0,1,2}$, and 6^1P_1 are solved to compute the density profiles of atoms in these states. The lowest eigenvalue of the ambipolar diffusion equation is computed, and from it the (constant) electron temperature, T_e. The corresponding eigenfunction is the electron density radial profile $n_e(r)$. The electrons are assumed to have a Maxwellian distribution, with a "cut off" due to the dc electric field, simulated by suppressing both direct ionisation and excitation to the resonance state 6^1P_1 from the ground state 6^1S_0. Radiation from the resonance states is included using the radially averaged diffusion coefficient approach of Holstein. [2] This approach has been shown to be in good agreement for our purposes with more exact Monte-Carlo simulations.[3] Transitions between all states listed above, together with ionisation from the excited states, are included in the diffusion equations. The neutral gas temperature profile is computed from the balance of heating by elastic electron collisions and thermal conduction.

In computing the contributions to the power balance in the discharge, excitation to the states 7^3S_1 and the triplet $6^3D_{1,2,3}$ are also included, but these states are assumed to radiate instantly on creation and the resulting radiation in the visible spectrum is assumed to be optically thin.

The code has successfully reproduced numerical results presented in [1] and also shows good agreement with experimental results of Verweij. [3] The paper will discuss the effects of noble gas excitation on these results, for a number of appropriate parameters.

References

[1] M.A.Cayless,"Theory of the Positive Column in Mercury-Rare Gas Discharges", PhD Thesis, University of London (1963),
M.A.Cayless, Proc 5th Int. Conf. on Ionization Phenomena in Gases, Munich, 262 (1961)

[2] T.Holstein, Phys.Rev., 72, 1212 (1947),
T.Holstein, Phys.Rev., 83, 1159 (1951)

[3] J.B.Anderson et al, Phys.Rev.A, 31, 2968 (1985)

[4] W.Verweij, Philips Research Reports Supplements, No 2 (1961)

Nonequilibrium thermodynamical discussion of ionisation in breakdown and glow

B. Nyíri
Tungsram Co. Ltd. Light Source Development Division
H-1340 Budapest, IV. Váci út 77. Hungary

It is shown that the ionisation equations derived by the methods of nonequilibrium thermodynamics for the case including higher electron drift velocities [1] can account for ionisation phenomena observable at various stages of discharge lamp operation; e.g. breakdown, glow, and local equilibrium.

These equations for ionisation rate are

$$r_1 = k_1 \cdot n_a - k_2 \cdot n_e n_i \cdot e^{-\frac{m_e v_e^2}{2RT}} \quad (1)$$

$$r_2 = L\, n_e \left[k_1 \cdot n_a - k_2 \cdot n_e n_i \cdot e^{-\frac{m_e v_e^2}{2RT}} \right] \quad (2)$$

for the normal and electron activated processes, respectively. The ratio of k_1 and k_2 is the equilibrium constant of the Saha equation (the velocity dependence of L is indefinite yet).

For breakdown/avalanche we solve

$$\frac{\partial n_e}{\partial t} + div(e\mu_e E n_e) = r_1 + r_2 \quad (3)$$

Its solutions in one dimension, with negligible space charge, and, in lack of space, considering n_i constant can be written as:

$$n_e = \frac{k_1 n_a (1 + L n_e^{\circ})\, e^{\lambda(\xi - \xi_o)} - (k_1 n_a - k_2 \cdot n_e^{\circ} n_i \cdot e^{-\beta E^2})}{(1 + L n_e^{\circ})\, k_2 n_i\, e^{-\beta E^2}\, e^{\lambda(\xi - \xi_o)} + L(k_1 n_a - k_2 \cdot n_e^{\circ} n_i \cdot e^{-\beta E^2})} \quad (4)$$

Here $\xi = x + vt$ with v being an arbitrary number, $n_e^{\circ} = n_e(\xi_o)$,

$\beta = \mu_e^2 m_e/(2RT)$, and

$$\lambda = \frac{k_2 n_i e^{-\beta E^2} + L\, k_1 n_a}{e\, \mu_e E + \upsilon} \tag{5}$$

If $\upsilon = 0$ equ.(4) describes an avalanche, while for $\upsilon > 0$ we have similar but non-stationary solutions. This avalanche is, however, saturated at a higher n_e depending on the field intensity (see the drift velocity υ_e dependence in equ.(1)).

To describe a glow-like situation we assume that the space charge is $e \cdot n_i$, thus also $n_e \simeq 0$. Since at low temperatures the r_1 ionisation rate reaches the saturation value at quite low electron kinetic energies i.e. at small E, we have in one dimension

$$e\mu_i \frac{d}{dx}\left(E \cdot n_i\right) = k_1 n_a \qquad \text{and} \qquad \frac{dE}{dx} = \frac{\varepsilon}{e} \cdot n_i \tag{6}$$

the solution of which is the 'hyperbola':

$$\left(E^2 - E_o^2 \right) = \frac{k_1 n_a \varepsilon}{e^2 \mu_i} \cdot x^2 + E_o \frac{\varepsilon}{e} n_i^o \cdot x \tag{7}$$

that has the asymptotes:

$$E = \pm \sqrt{\frac{k_1 n_a \varepsilon}{e^2 \mu_i}} \left[x + \frac{E_o e\, \mu_i n_i^o}{2\, k_1 n_a} \right] \tag{8}$$

For finer analysis of the above results the inclusion of proper boundary conditions [2] and of diffusion is necessary.

[1] *Nyiri, B.: Proceedings of the symposium on High Temperature Lamp Chemistry II. of the Electrochemical Society in 1988. (El. Chem. Soc. Inc.) Proc. Vol. 88-4 pp.170.*

[2] *pp. 185. in* [1]

DEUTERIUM LAMP WITH IMPROVED LIFE TIME

W. Schwarz, P. March

W.C. Heraeus GmbH, Produktbereich Original Hanau,
D-6450 Hanau

Deuterium lamps emit a continuous spetrum in the wavelength region of about 180 ... 350 nm. They are well established as light sources in spectrophotometers and for UV-calibration purposes (transfer standards).

Their life time is defined to be the operation time with a decrease of 50 % of the initial intensity. Because of the high transmission in the spectral region of interest, quartz glass is a prefered material for the lamp bulb. Such lamps show a life time of about 500 to 750 hours in the region below 250 nm.

We will present a new lamp type with an improved life time up to 2000 hours.

The intensity decrease is mainly caused by the evaporation of cathode material, especially earth-alkalies and the following deposition at the inner wall of the quartz-envelop. The deposited earth-alkalies show reactions with the SiO_2-network which lead to transmission loss especially in the above mentioned spectral region. The improvement in life time could be achieved by applying a protective coating which reduces the attack of earth-alkalies.

DEUTERIUM LAMP FOR SPECTRO-ANALYTICAL DEVICES WITH REDUCED NOISE LEVEL

W. Schwarz, P. March

W.C. Heraeus GmbH, Produktbereich Original Hanau,
D-6450 Hanau

Deuterium lamps are high sophisticated DC-driven light sources emitting continuous radiation in the far UV with a high constancy. This DC-light level is superimposed by a stochastic noise with a level of about 10^{-4}. However, some applications require a level of about 10^{-5}. Our investigations have shown, that the noise is caused by the formation of ozon. The lamps radiation at a wavelength of about 185 nm is absorbed by air. This absorption leads to the formation of ozon, which itself absorbes in the region 200 ... 300 nm. Fluctuations in the ozon concentration across the whole light path leed to fluctuations in the light signal. - This is the origin of noise. To reduce the noise level one has to prevent the ozon formation. This is achieved by applying a dielectric filter-coating at the outer wall-side. This coating shows a steep transmission edge with a 50-percent-value at about 195 nm. The transmission at 200 nm is 70 % at least, so that the transmission of the quartz-glass in the spectral region above 200 nm will be almost not affected. Our investigations have shown, that the transmission has to be less than 30 % in the critical region nearly 185 nm to prevent ozon formation.

CALCULATIONS OF FLUORESCENT LAMP PARAMETERS

R. Sonnenburg, H. Gündel*

VEB Kombinat NARVA, ZFT, DDR 1017 Berlin, Ehrenbergstr.11-14

*Zentralinstitut für Elektronenphysik, DDR, 1086 Berlin, Hausvogteiplatz 5-7

At the Karlsruhe Symposium we reported a method of computermodelling of fluorescent lamp plasmas which allows the evaluation of lamp parameters without any assumptions concerning the lamp current. In this paper we present some results of calculations performed on the basis of a slightly modified model compared to that given in the paper [1]. The modifications concern the number of energy levels of the Hg-atoms included and the way how the deviations of the electron energy distribution function from a Maxwell like form are involved. Here we are computing both the 253,7 nm and the 184,9 nm line output. For taking into account the depletion of electrons in the high energy tail of the distribution function from results of Winkler et al.[2] an approximation of the following kind was made $T_t = T_b - f(E(t) * I(t))$ with T_t-tail temperature, T_b-bulk temperature of the electrons, $E(t)$-electric field strength, $I(t)$-lamp current and t-time. The function f has been taken instead of the factor 0,4 used in [1,3] for fitting the distribution function. All calculations reported here were performed for a tube radius range from 4 mm to 18 mm at a supply frequency of 50 Hz and a gas temperature of 42°C. It has been calculated the current dependence of the electric field strength, of the number densities of all species included, of the efficacy of the radiation of both lines and of the elastic and inelastic losses. Fig. 1 shows the corresponding graphic for the electric field strength with the expected behaviour.

In order to compare our results with those from Bouwknegt [4] who obtained an optimal value of the electric field strength for lamp efficacy we determined the dependence of the various kinds of energy dissipation in the lamp plasma on the field strength at fixed current 200 mA (Fig. 2). For the efficiency of the 253,7 nm line radiation output we also find a slight maximum at about 1,2 V/cm. Quantitative agreement of our model with experimental values has to be improved.

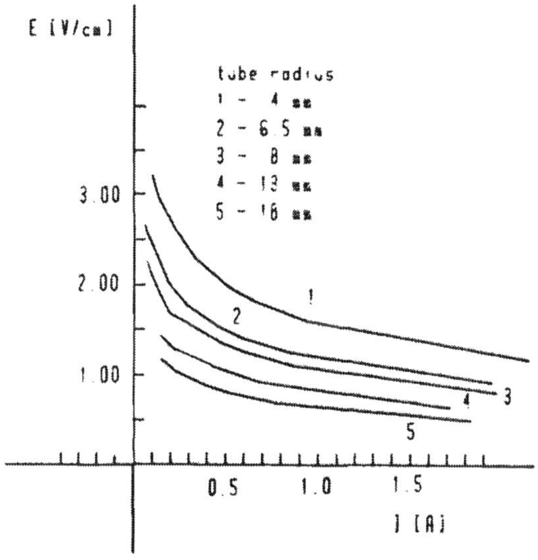

Fig. 1

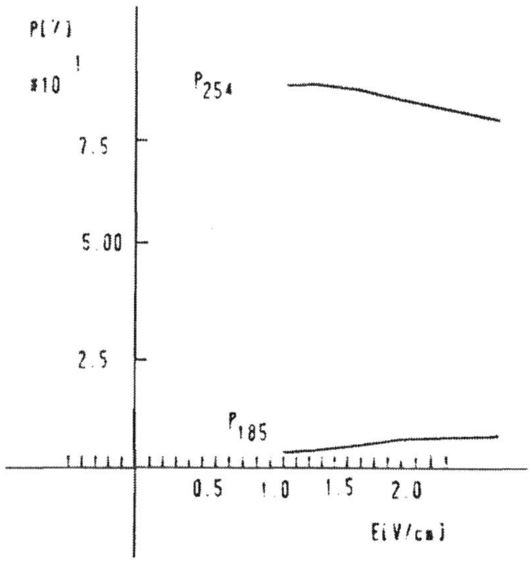

Fig. 2

[1] Sonnenburg,R., Gundel,H.: Contrib.Plasma Phys.27(1987)1,37

[2] Winkler,R.B. et al.: Ann. Physik 40(1983)90

[3] Polman.J. et al.: J.Phys.Lond. D: Appl.Phys. 5(1972)266

[4] Bouwknegt,A.: J.IES 11(1982)204

PRELIMINARY REPORT ON THE NOVEL TECHNOLOGY OF CIRCULAR FLUORESCENT LAMP

Mi Zhongliang ,

Director of R & D ,

Shanghai No.2 Special Lamp Works ,

3500 North Zhong Shan Road ,

Shanghai , China

The circular fluorescent lamps have become more and more popular now.

A novel processing of bending circular fluorescent lamp, quite different from conventional one , will be discribed. This new technology will bring about advantages , such as low-cost of equipment , easy-control of operation , high-quality of product .

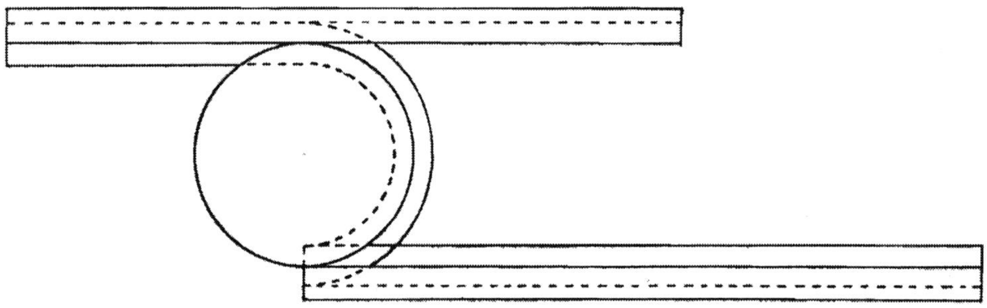

On the experimental and theoretical determination of the electric field strength for the super-narrow tube diameter, low pressure Hg/Ar lamps.

G.Zissis, P.Bénétruy, J.J.Damelincourt
Laboratoire de Décharges dans les Gaz (URA-277 CNRS)
118, rte de Narbonne, 31062 Toulouse, FRANCE
A.Por
Cie Philips Eclairage, Pont-à-Mousson, FRANCE

The low pressure Hg+rare gaz discharge is of a great interest because of its practical applications (i.e. fluorescent lamps). During the last years the commercial importance of the Hg/Ar lamps with super-narrow tube diameter (<3mm) has progressed considerably because of the numerous special applications of this kind of lamps.

Early experimental investigations (1) of the positive column of a low pressure Hg/Ar lamp with narrow tube diameter have shown that the E-I characteristic of the discharge is different from that of the classical cases (δ>20mm). For such discharges, the numerical results from most of the existing classical models (2, 3) disagree with the observed experimental values. This is particularly true for the electric field value, because these models are based on the classical considerations concerning the Electron Energy Distribution Function (EEDF) and radiation escape from the plasma. This discrepancy is accentuated as the cold spot temperature becomes lower.

This paper summarizes the first steps of a more extensive study of such discharges and thus we deal with measurements and calculations of the electric field in the positive column.

On one hand, electrical measurements have been carried out for several hot cathode Hg/Ar lamps with 1.5, 3 and 6mm tube diameter, operating with a high frequency (28 kHz max.) a.c. power supply. The Ar pressure was 15 Torr (24°C) in all cases. The cold spot temperature has been fixed within ±2°C by submerging the tube in a bath regulated with oil circulation. Measurements has been carried out in the range of -15 to +85°C for two different currents (20 and 40 mA). Electric field strength has been deduced from the arc voltage measurements. Electrode potential fall was previously determined from voltage measurements for several tube lengths (15 and 22 cm). The maximum experimental error on the values of electric field has been estimated to be ±6% in all cases.

On the other hand, we have developed a self-consistent model. In this model, we deal with calculations of the population densities for the fundamental and the 5 first excited Hg levels, electron and rare gas temperature and densities. All these quantities have been carried out as functions of the radial position in the positive column, which has been considered to be axially uniform. Electric field and radiative power output have also been calculated. An iterative procedure has been used in order to solve simultaneously the 6 continuity equations for the Hg atomic levels, the ionization balance equation and the energy conservation equations for the neutral gas and the electron gaz. In addition local charge neutrality, electric current equation and perfect's gas law have been introduced in order to complete the equation set.

In our calculations we use a non-Maxwellian EEDF as given by Lagushenko (4), because it was reported (5) that this analytical function reproduced better experimental results than the Two Electron Group approximation for the EEDF used frequently in the literature (6). Diffusion controlled positive column has been assumed and volume recombination has been neglected. Diffusion of the metastable Hg levels was taken into account. It is important to note that all transport coefficient and reaction rates have been calculated at each step as functions of the radial position. The model does not contain any adjustable parameter and the cross sections are taken from the literature (7). Holstein's (8) theory for the two Hg resonance lines imprisonment was used as a first approximation. The Ar influence was also taken into account by using the Walsh (9) theory.

Figure 1 summarizes our experimental and theoretical results for the electric field as a function of the cold spot temperature in the case of the 1.5mm, 20mA Hg/Ar lamp. The results of the classical model proposed by Waymouth are also reported on the same figure. We may conclude that our model reproduces much better experimental values than the classical model. We may also observe some important discrepancies between experimental and theoretical values in the case of lower cold spot temperatures. This may be due to the argon contribution to the ionization in the discharge. This latter conclusion is based on the high values of the electron temperature obtained by both classical (>100000K) and our model (40000K), and the important values of cross-sections for the stepwise ionization of Ar. An additional experimental confirmation was the observation of a violet halo emitted by the discharge under these conditions, this halo may be characteristic of the Ar excitation.

Figure 1 : Electric field as afunction of the cold spot temperature for a lamp with δ=1.5mm and I=20mA.

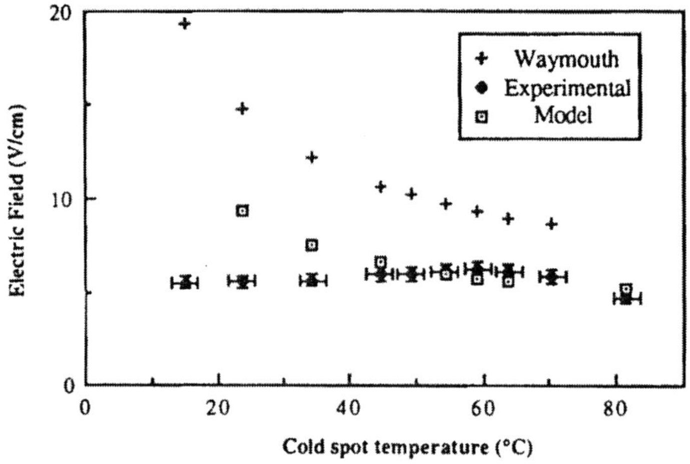

(1) Koedam, M., Kruithof, A.A., Physica, 28, 80, 1962.

(2) Cayless, M.A., Brit. J. Appl. Phys., 14, 863, 1963.

(3) Waymouth, J.F., Electric Discharge Lamps, M.I.T. Press, Cambridge, 1971.

(4) Lagushenko, R., Maya, J., Journal of IES, 14, 306, 1984.

(5) Zissis, G., Yousfi, M. et al, XIX-I.C.P.I.C., Belgrade, 1989 (to be published).

(6) Dakin, J.T., J. Appl. Phys., 60, 563, 1986.

(7)Winkler, R.B., Wilhelm, J., Winkler, R., Ann. Phys., 40, 89,1983.

(8) Holstein, T., Phys. Rev., 72, 1212, 1947; 83, 1159, 1951.

(9) Walsh, P.J., Phys. Rev., 116, 511, 1959.

Comments on reaction rate calculations in positive column of low pressure Hg-Ar discharges

G. Zissis, A. Alkaa, M. Yousfi and J.J. Damelincourt

Université Paul Sabatier
Laboratoire de décharge dans les Gaz URA 277, CPAT
118, Route de Narbonne, 31 062 Toulouse Cedex (FRANCE)

It is well known that knowledge of reaction rates of various excited Hg states by electron impact is essential for modeling of low-pressure Hg-Ar discharge lamps. Reaction rates can obviously be obtained from experiments, however, due to the very insufficient experimental data available in the literature, they are generally calculated using the collision cross sections and the electron energy distribution function. Numerous studies are already devoted to the determination of electron energy distribution but results are not always consistent. In a recent comparative study[1] of electron energy distribution in positive column of low pressure Hg-Ar discharge, the authors have already shown non negligible deviations between distribution functions obtained from five different methods, more particularly in the distribution tail (see Fig. 1 hereafter). Therefore, the aim of this communication is to emphasize the importance of the choice of calculation method of distribution function on the evaluation of Hg reaction rates.

Fig. 1 shows the different electron energy distributions in Hg-Ar mixture used for the calculation of Hg reaction rates. The discharge conditions are : tube diameter 36 mm, 3 Torr of Ar (20 °C), 40°C for wall temperature (i.e. 6.304 mTorr of saturated Hg vapor) and 400mA dc current. We observe on Fig. 1 a relatively good agreement between the various distribution function in the bulk energy region i.e. for electron energy lower than ~ 4 eV. However, for higher electron energy i.e. for the distribution tail, more or less important discrepancies appear. Maxwell distribution at temperature T_e=11324 K is the upper limit and the two-electron group distribution (with a bulk temperature of 11324 K and a tail temperature of 6000 K) is the lower limit. We also observe on Fig. 1, up to 7 eV, a good consistency between Allis' distribution fitted in accordance to Damelincourt[2]'s probe measurements and Lagushenko[3]'s distribution. Furthermore, the Lagushenko's distribution, which is an analytical approximate solution of Boltzmann equation, presents an interesting coherence with our Boltzmann calculation. Our two-term method of solution of Boltzmann equation (including elastic, inelastic and super-elastic electron-atom collisions and also electron-electron and electron-ion interactions) is already described elsewhere[1].

As an example of our calculation, we report in the following table, the relative difference between some Hg reaction rates calculated with the help of Maxwell distribution (chosen as a reference because the Maxwell distribution is easier to reproduce) and compared to the reaction rates calculated from the other distribution functions shown on Fig. 1. These reaction rates correspond to the electron impact excitation of the 6^1S_0 Hg level.

Excited States→ References ↓	6^3P_0 %	6^3P_1 %	6^3P_2 %	6^1P_1 %	7^3S_1 %
Lagushenko's relation	25	15	75	84	96
Allis-Damelincourt	34	21	99	99.9	100
Two electron group model	53	45	92	95	99
Boltzmann calculation	53	47	83	87	96

We observe in this table a strong dependence between the relative difference of reaction rates and the distribution function chosen. This relative difference can reach 90 % in the case 7^3S_1 Hg rate obtained from the two group distribution. For a given process, in order to understand the disparity of relative difference as a function of energy distribution, it is interesting to establish a parallel between the energy threshold and the magnitude of Hg excitation cross section on the one hand and the energy position of the distribution tail considered on the other hand, because the reaction rate values depend essentially on the intersection area between distribution tail and excitation cross section. For this reason, we observe a better agreement between the reaction rates calculated from Lagushenko's relation and from our numerical Boltzmann method for the excitation processes with a higher energy threshold (i.e. 6^3P_2, 6^1P_1 and 7^3S_1) because the tails of these two distribution present a better agreement for higher energy region (see Fig. 1). Furthermore, we show on Fig. 2 the relative difference between reaction rates calculated from Maxwell distribution and other distributions ; curves are plotted as a function of energy level. The positive energy region corresponds to the direct transition between the 6^3P_0, 6^3P_1, 6^3P_2, 6^1P_1 and 7^3S_1 levels due to electron impact and the negative energy region corresponds to the inverse processes i.e. to the super-elastic collisions. Fig. 2 clearly shows a weak dependence of reaction rates (lower than ± 5 %) on the energy distribution in the case of super-elastic processes, because the magnitude and the threshold of the corresponding cross sections are lower than the direct excitation ones. However, this dependence in the case of the direct excitation processes generally increases with the energy threshold.

In conclusion, the choice of the distribution function used for reaction rate calculations in low pressure Hg-Ar column positive discharge must be carefully considered, because differences between the various reaction rates obtained are far to be negligible, especially in the case of the reaction rates corresponding to the direct electron-Hg atom impact excitation. Normally, the more precise distribution function should correspond to the Boltzmann equation solution. However, in order to improve our numerical Boltzmann solution, we needs to include the treatment of Penning reactions which play a relatively important role in low pressure Hg-Ar discharge.

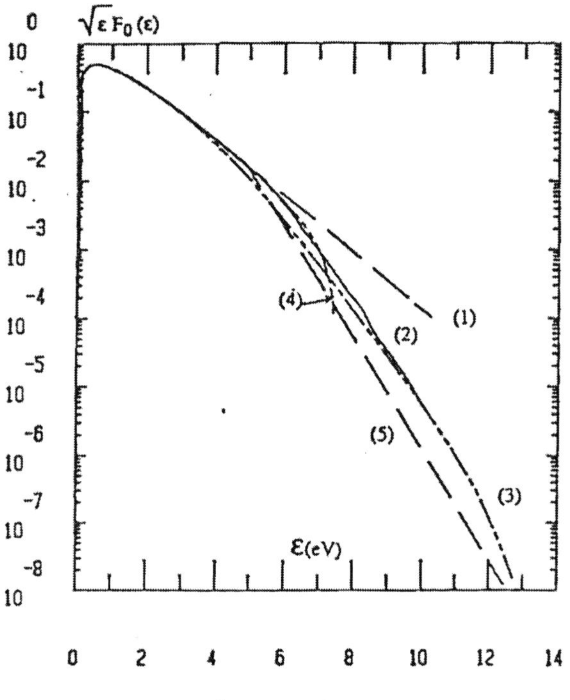

<u>Figure 1</u> : Electron energy distribution in Hg-Ar positive column discharge. T_w= 40 °C, i=400 mA, r=18 mm, p_{Ar} = 3 Torr at 20 °C :
(1)Maxwell distribution (2) Lagushenko's relation
(3)Boltzmann calculations
(4) Allis-Damelincourt distribution
(5) Two electron group model

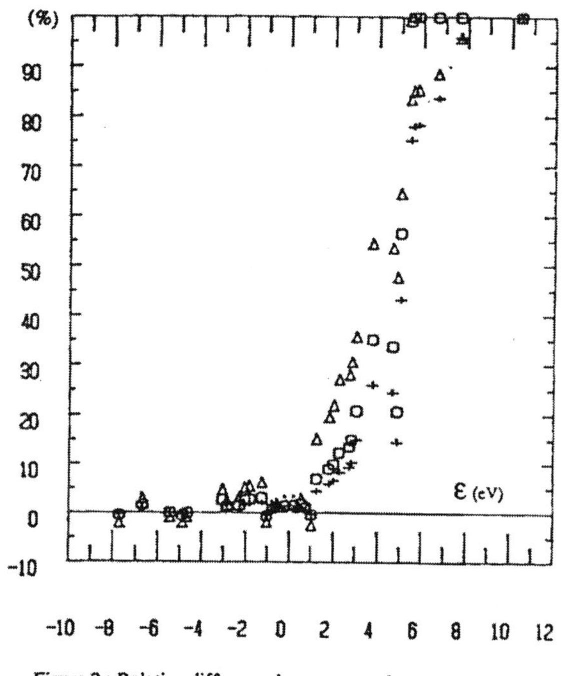

<u>Figure 2</u> : Relative difference between reaction rates calculated from Maxwell distribution and from Lagushenko expression (O O O), Allis-Damelincourt distribution (+ + +), Boltzmann equation calculations (Δ Δ Δ), as a function energy threshold

Acknowledgement : J. J. Damelincourt and G. Zissis wish to thank Cie Philips Eclairage (Pont-à-Mousson) for partial support, while M. Yousfi and A. Alkaa wish to thank the members of the European Group on Gas Discharge for numerous helpful discussions.

[1] G. Zissis, A. Alkaa, M. Yousfi and J.J. Damelincourt, ICPIG IXX, (1989) Belgrade
[2] J. J. Damelincourt, "Thèse de Doctorat d'Etat", n° 566, Université Paul Sabatier, (1973), Toulouse
[3] R. Lagushenko and Maya J., J. of IES, (1984), 306-314

ADDITIONAL ELECTRODE HEATING DURING DIMMING OF HF OPERATED FLUORESCENT LAMPS.

F.A.S. LIGTHART, H. TER HEYDEN AND L. KASTELEIN.

PHILIPS LIGHTING B.V.

EINDHOVEN, THE NETHERLANDS

With the advent of electronic ballasts for fluorescent lamps power regulation has become more and more easy and attractive. In addition to HF-operation at nominal power there is an increasing demand for HF-dimming until even a 10% power level.

At low power levels of the fluorescent lamp the lamp current is insufficient to keep alive the hot spot of the electrode. Additional heating of the electrode is necessary. At low lamp current and low additional heating currents the electrode temperature is too low for thermal emission and as a result sputtering occurs, while at high additional heating current and/or high lamp currents fast vaporization of electrode material occurs. Therefore, two problem areas can be discerned: A sputter area at low lamp current and/or low additional heating current and a vaporization area at high lamp current and/or high additional heating current.

When the lamp operates in one of these problem areas early end-blackening occurs and possibly early failure. Between these two problem areas is situated a preferred regime of lamp and heating currents leading to good lamp behaviour.

The boundaries of additional heating have been studied for the 5ft T8 argonfilled TLD 50W/HF and for the compact fluorescent PL 11 W/HF.

As a criterium for the sputter and the vaporization areas is used end-blackening within 1000 hours of operation. For TLD 50W/HF it appears that additional heating is necessary below a lamp current of 250 mA. Above an additional heating current of about 430 mA vaporization occurs, even at very low lamp currents. At low dimming levels only a narrow range of additional heating currents are admitted in order to preserve good lamp behaviour. See enclosed figure.

In addition to the experiments a theoretical expression has been derived for the sputter curve, which forms the boundary of the sputter area.

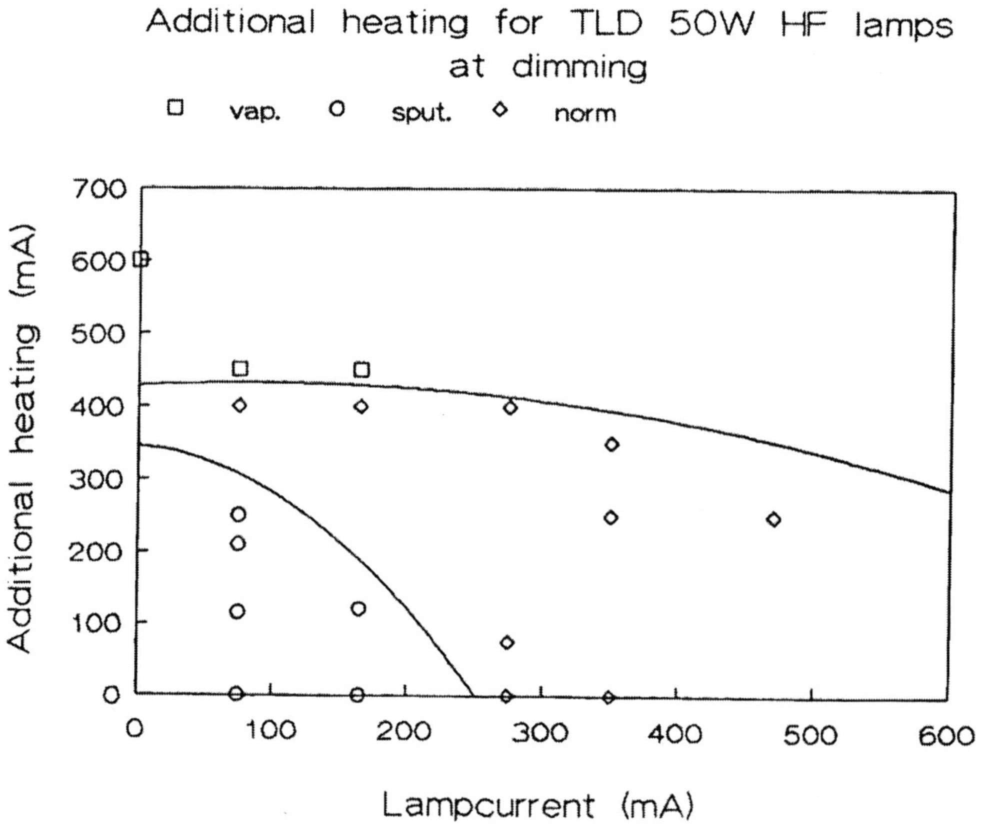

SUSPENSION FLOW-COATING OF FLUORESCENT LAMPS: MODELLING OF THE DRYING STEP

F.W. Altena*, M. Westerhof**, A. Boomkens***
and M. van Rijswick***

* Philips Lighting B.V., Eindhoven, The Netherlands
** Center for Manufacturing Technology, Nederlandse Philips Bedrijven B.V., Eindhoven, The Netherlands
*** Philips Lighting B.V., Roosendaal, The Netherlands

ABSTRACT

Control of fluorescent powder layer coating in fluorescent lamps is important for reasons of optimal luminous flux and chromaticy coordinate control. Translated into powder layer properties this means that properties such as layer thickness and morphology (e.g. particle packing, layer texture) should be well defined.

This paper adresses one aspect i.e. the problem of controlling the powder layer thickness distribution in a fluorescent lamp from a modelling point of view. After the initial vertical coating of a fluorescent bulb with a suspension containing the desired phosphor, the solvent is removed in a convective drying procedure, where drying air is flushed through the wetted bulb. The model proposed in this paper aims to described this drying step of the phosphor layer, which is critical to obtaining optimal fluorescent lamp properties.

The drying step is a rather complicated process involving interlinked heat and mass-transfer (evaporation) and suspension rheology under conditions where local temperature and layer composition may vary considerably. In the model, the flow and temperature fields have been solved, and mass transport through evaporation has been accounted for. Supported by experiments, it is assumed that the evaporation rates of solvent from suspensions or drying powder layers equals the rate for pure water. A simple power-law fluid flow model appears to adequately describe the suspension flow. By combining relatively simple sub-models, a complete model description of the drying process can be obtained, without adjustable parameters.

The model results are very sensitive to details of the suspension rheology. The flow behaviour of a drying suspension has been simulated through viscosity measurements on suspensions where the solvent content has been removed in a way corresponding to the actual drying conditions.

Model predictions are compared with coating experiments where both suspension composition and drying conditions have been varied. As a model system we took a calcium halophosphate phosphor in a suspension containing carboxymethylcellulose as a binder. The imposed conditions of the drying process are kept constant within one coating experiment. Experimental and calculated powder layer distributions are compared, and also other characteristics such as total required drying time and temperature distributions in the drying bulb have been compared to judge the validity of the model. The quantitative agreement appears to be fair. Qualitatively, the effect of varying some critical parameters on the outcome of the drying process can adequately be estimated.

Some attention is paid to application to a drying process where during drying, parameters are varied such as drying air flow rate and temperature. This situation represents the more or less practical drying conditions aimed at achieving an optimal powder layer distribution along the fluorescent bulb. In view of the model complexity and underlying assumptions as yet to be tested the modelling efforts are judged to be encouraging and may help to understand and improve present-day coating operations.

Review of Compact Fluorescent Lamp R&D

Seiichi Murayama, Central Research Laboratory, Hitachi, Ltd.
Kokubunji, Tokyo 185, Japan.

1. Introduction

Demands for energy saving increased, after the oil crisis in 1973. Compact fluorescent lamps were intensively studied to replace incandescent lamps. Many technical advances have been made since that time, and many types of compact fluorescent lamps have been developed. This paper reviews technical progress in compact fluorescent lamps.

2. Old ideas on compact fluorescent lamps

It is only about ten years since the first compact fluorescent lamp(1) became commercially available. This was a circular lamp with a resistance ballast, a starter, and an Edison base. However, some ideas with respect to compact fluorescent lamps were born in the early stages of the fluorescent lamp development. Researchers naturally desired to make a lamp with a similar shape to an incandescent lamp, utilizing the new principle, that is, the fluorescence of phosphor excited by low-pressure mercury-vapour discharges.

The following are some examples of old patents. In 1938,

Campbell(2) invented a lamp with partitions to force the discharge to occur along a path much longer than the straight line joining the electrodes. In 1939, Henninger, Jr. and Henninger(3) invented a fluorescent lamp with a helical tube, which served a concentrated light source. In 1946, Fujiki(4) invented a lamp with two U-shaped discharge tubes, a common reactance ballast, and an Edison-type base.

3. Prototypes of single-ended fluorescent lamps

Although many ideas with respect to compact fluorescent lamps were born in very old times, as stated above, prototypes of compact fluorescent lamps were first presented in scientific papers in the 1970s. Campbell(5) described a compact fluorescent lamp comprising an inner tube surrounded by an outer tube, thus forming an annular construction. Multi-channeled arcs were produced time-sequentially in the annular space by switching circuits. Roche(6) reported a compact fluorescent lamp with an igniter coil in the discharge tube for auto-ignition. Hasker(7) introduced into the positive column a recombination structure consisting of glass wool thinly distributed over the dischage volume; recombination loss of electron-ion pairs on the surface of the recombination structure increased the field stregth and the efficacy. Gross and Skeist(8) applied a magnetic arc-spreading technique to a compact fluorescent lamp. Yamane and Watanabe(9) discovered a self-switching action caused by anode oscillation in a multi-arc lamp with an annular construction, with a cathode in the inner tube and multi-anodes the annular space. The lamp was operated by a single ballast connected commonly to the multi-anodes. Young and Morton(10) examined compact fluorescent lamps with partitions to keep positive column length long enough to obtain high efficacy. Yamane and Watanabe(11) found that the positive column in a transverse magnetic field moved in the retrograde direction through the annular space between coaxial glass tubes; the experiments were made with a lamp structure applicable to a compact fluorescent lamp. Anderson(12) proposed a system in which an electrodeless discgarge was produced in a toroidal vessel. This idea developed into a system which had a geometry more suitable for a compact fluorescent lamp.(13,14)

4. Technical progresses in compact fluorescent lamps

Efficacy of a fluorescent lamp, η, is given by

$$\eta = (\eta_A V_A + \eta_P V_P + \eta_C V_C)/(V_A + V_P + V_C) \quad (1)$$

where η_A, η_P, and η_C are efficacies in the anode fall, the positive column, and the cathode fall, respectively. They depend on the conditions of lamp structure, phosphor material, gas composition, current, mercury vapour density and so on. V_A, V_P, and V_C are voltages of the anode fall, the positive column, and the cathode fall, respectively. V_P depends on various discharge conditions, while V_A and V_C do not depend so much on the discharge conditions in normal types of fluorescent lamps. η_A($\simeq$ 0 lm/W) and η_C($\simeq$ 30 lm/W(15)) are smaller than η_P($\simeq$ 80 lm/W). Therefore, η increases with V_P.

4.1 Method of increasing

There are two ways to increase V_P: one is to increase the length of the positive column, and the other is to increase its field strength. To increase the positive column length in the restricted volume of a compact lamp, it is nescesary to form a zigzag pathway. Recent advances in glass technology have removed the difficulty in forming the zigzag pathway. The simplest and easiest way is to bend the tube or connect two straight tubes by a small bridge through which discharge passes.(16)

The simplest and best way to increase the positive-column field-strength is to decrease the diameter of the positive column. However, this method creates new problems.

These will be discussed in the following sections.

4.2 Mercury atom density control

Reducing the diameter decreases the lamp surface area and increases the coldest-spot temperature, thus increasing the mercury atom density to a much higher value than is suitable for obtaining good efficacy. There are three methods of controlling the mercury atom density. The first is to create cool spots in the discharge tube, at a temperature at which the saturated vapour pressure of mercury is appropriate for obtaining good efficacy. In a lamp structure in which two straight discharge tubes are connected by a small bridge,(16) it is rather easy to create cool spots. If the bridge is positioned a small distance from the tube end, current does not flow in the end portion. Thus, the temperature of the end portion is lower than that of the part through which current flows. The temperature of the tube end is controlled by the distance between the bridge and the tube end. Where the discharge tube is bent in a U-shape, cool spots can be provided by making the radius of the bent part larger than that of the straight part. In the structure in which the discharge tube is covered with a light-diffusing cover, holes and/or slits are formed in the light diffusing cover to cool the discharge tube by ventilating.(17) It is also effective to fill with silicon oil the space between the outer cover and the discharge tube, thus, cooling the discharge tube by thermal conduction.(18)

The second method is to use amalgam. If the composition of mercury alloys is well selected, the mercury vapour pressure can be kept near an optimum value for obtaining good efficacy, over a wide range of temperature.(19) One example of a mercury alloy is Bi-In-Hg.

The third method is to set discharge tubes with one open end inside an outer bulb, thus, allowing free access of gas from the outer bulb.(20,21,22) Since the coldest spot is formed on the surface of the cooler outer bulb and not on the hotter discharge tubes, the vapour pressure of mercury is kept at a suitable value. Besides, the density of mercury atoms in the inner tube is much smaller than that in the outer bulb, because the temperature of the inner tube is much higher than that of the outer bulb.(20,21)

4.3 Improvement in lumen maintenance

The smaller diameter causes an increase in wall loading and intensity of ultra-violet radiation (Hg 185 nm)(23), causing a severe deterioration in phosphor conversion efficiency.(24) Recently, efficient rare-earth phosphors have been developed, which have narrow spectral emission bands.(25) These phosphors have high stability in intense ultraviolet radiation.(24) Thus, they enable realization of a compact fluorescent lamp with a long life and a high luminous efficacy, together with a high colour rendering index, as was theoretically predicted(26,27).

Increase in wall loading increases the rate of glass darkening due to penetration of mercury into the glass(28). This can be weakened by coating a metal oxide layer on the surface of the glass(21,29,30).

5. Possible Future Progress

High frequency operation using semiconductor circuits improves lamp system efficacy because of the absence of anode-fall loss(31) and the decrease in power loss in the ballast. Semiconductor circuits also have advantages of lightness, and ease of instantaneous starting and dimming. Unfortunately, at the present time, the use of semiconductor circuits is limited by their high cost and radio-frequency noise. However, the potential of high frequency operation using semiconductor circuits is expected to cancel out the disadvantages, especially in compact fluorescent lamps. That is, the radiation of radio-frequency noise is less severe in the compact fluorescent lamp because the U-shaped discharge tubes form parallel discharge paths in which the discharge currents flow in opposite directions, thus, cancelling the electro-magnetic field. Besides, IC technology is expected to advance further. Use of an IC circuit may decrease the volume of ballast and make it easy to realize complicated lamp operations which will become common in future lighting systems.

5. Conclusion

According to the preceding discussions, technical advances in phosphor and glass technology have made many kinds of compact fluorescent lamps comercially available. High frequency operation of compact fluorescent lamps using IC circuits is expected to be more widely used in future lighting systems.

6. References

1. Hirota,T., Ohno,H., Hosoya,K., and Ohta,S., J. Light & Vis. Env., 3,1(1979).
2. Campbell,N.R., British Patent 513715(1938).
3. Henninger,Jr.,A.F. and Henninger,G.D., US Patent 2281579 (1939).
4. Fujiki,U., Japan Utility Model 363115(1946).
5. Campbell,J.H., Journal of IES, 2,3(1972).
6. Roche,W.J., Journal of IES, 3,65(1973).

7. Hasker,J., Journal of IES, 5,29(1976).

8. Gross,L. and Skeist,S.M., LD & A, 9,31(1979).

9. Yamane,M. and Watanabe,Y., J. Illum. Engng. Inst. Jpn., 63,549(1979).

10. Young,R.G. and Morton,E.W., LD & A, 10,38(1980).

11. Yamane,M. and Watanabe,Y., J. Appl. Phys., 56,746(1984).

12. Anderson,J.M., Illum. Eng., 64,236(1969).

13. Anderson,J.M., US Patent 3521120(1968).

14. Hollister,D.D., US Patent 4010400(1975).

15. Matsuno,H. and Murayama,S., Proc. Annual Meeting of Illum. Engng. Inst. Jpn. in 1988, p.7.

16. Bouwknegt,A., Journal of IES, 11,204(1982).

17. Kamei,T., Hayashi,M., Osada,K., Ikeda,S., and Takano,Y., Toshiba Review, 35,679(1980).

18. Sangen,M. and Arashima,M., Proc. 16th Annual Meeting of Illum. Engng. Inst. Jpn. in 1983, p.16.

19. Bloem,J., Bouwknegt,A., and Wesselink,G.A., Journal of IES, 6,141(1977).

20. Ono,T. and Matsuno,H., J. Illum. Engng. Inst. Jpn., 68, 254(1984).

21. Murayama,S., Matsuno,H., Watanabe,Y., Ono,T., Hosoya,K., and Hirota,T., Journal of IES, 14,298(1984).

22. Yamasaki,H., Takeda,T., Yamamoto,K., Nakano,S., and Tanaka,N., MITSUBISHI DENKI GIHO, 58,792(1984).

23. Denneman,J.W., De Groot,J.J., Jack,A.G., and Ligthart,F.A. S., Journal of IES, 10,2(1980).

24. Vrenken,L.E., Lighting Res. & Technol., 10,161(1978).

25. Verstegen, J.M.P.J., J. Electrochem. Soc., 121,1023(1974).

26. Thornton,W.A., J.O.S.A., 61,1155(1971).

27. Koedam,M. and Opstelten,J.J., Lighting Res. & Technol., 3, 205(1971).

28. Mulder,B.J. and Van Heusden,S., J. Electrochem. Soc., 130, 440(1983).

29. Ozono,T., Nishikage,Y., and Urata,K., Proc. 18th Annual Meeting of Illum. Engng. Inst. Jpn. in 1985, p.12.

30. Matsuno,H., Murayama,S., Ono,T., Kuga,K., and Torii,I., Proc. 19th Annual Meeting of Illum. Engng. Inst. Jpn. in 1986, p.15.

31. Koedam,M. and Verweij,W., Proc. 7th Int. Conf. on Phenomena in Ionized gasses, Belgrade, 1965,3,392-395.

CHEMICAL AND PHYSICAL MODELLING OF METAL HALIDE DISCHARGES - TODAY AND TOMORROW

E.Fischer and E.Schnedler
Philips GmbH Forschungslaboratorium Aachen
D-5100 Aachen, Federal Republic of Germany

1. INTRODUCTION

Talking on modelling of metal halide discharges does not simply mean to present methods that describe the properties of an arc plasma with given composition, but it is the attempt to mediate a feeling for the various phenomena that contribute to the behaviour of a very complex system. This system consists of several very different parts, which all influence each other by physical or chemical interactions.

Four main parts should be discerned that need different methods of modelling: the arc tube, the electrodes, the salt melt and the arc plasma. This is already a simplification, as also the outer electrical circuit and the heat balance of the luminaire may be important in some aspects.

Interactions between the parts take place on very different time scales. Processes within the plasma and phenomena at the electrode surfaces or changes in the electrical circuit take place within milliseconds or even microseconds. Much longer than this "plasma time scale" are the typical times in which the thermal properties of the solid or liquid parts of the system may change. The "thermal time scale", the time in which the system reaches a stationary state after a change of some outer condition is in the order of seconds or minutes. The third, the "chemical time scale", is the one which is the most important for lamp manufacturers. It covers all the corrosion or loss processes that change the lamp properties on a scale of hundreds or thousands of hours and finally limits the useful life time of all lamps.

One might think of a complete lamp model that covers all the phenomena in the complex system on all the time scales, but such a model does not exist today as far as we know and it is questionable if it would ever pay off the effort that is necessary to set it up.

Here we will focus our interest on existing models for the various parts of discharge lamps, first on models that can be regarded as stationary or quasi-stationary on the thermal scale, models that describe the thermal, electrical and radiative properties of a lamp at a certain point of time, and then on modelling of the transport processes and changes of lamp data that take place on the chemical time scale.

2. INTERACTIONS IN THE LAMP SYSTEM

Let us first look on the four parts of a metal halide discharge lamp. What are the properties we are interested in for any of the subsystems and how are they connected to one another on the different time scales?

At the first glance the arc tube is the part of the lamp that can be described most easily. Its temperature profile may be determined, if we know the heat flux from the arc plasma and from the electrodes. But on the chemical time scale we also must consider reactions between the tube material and the salt filling, diffusion of parts of this filling through the wall (e.g. sodium or hydrogen), changes in the tube energy balance due to wall blackening and release of impurities contained in the wall material.

Apart from the fact that in real lamps the electrode geometry is rather complicated, modelling of electrodes would only consist of solving the heat flux equation with radiative boundary conditions, if the energy input from the plasma was known. But determining this input flux already requires some knowledge of the plasma properties in the boundary layer at the electrode surface. Still more details are required to calculate corrosion rates and geometrical changes during lamp life.

Primarily of course we are interested in the properties of the lamp plasma that determine the light output and the colour rendering properties of the lamp. These data are not directly related to the filling charge. As the largest fraction of the metal halides in not in the gas phase but remains as a melt near the coldest spot of the tube wall, the pasma properties are strongly coupled to the temperature of the tube and to the chemical processes that occur in the complex salt melt.

So all the components of a lamp are mutually connected and modelling of one part is not possible without making at least reasonable estimations on the interaction with the others.

Today various models exist regarding indivdual aspects of metal halide lamps and first attempts have been made to couple them at least for the description of the stationary state of the lamp.In the next sections we will first discuss some of these models, then some models concerning transport and corrosion phenomena will be presented and finally we will discuss the question, which phenomena are not yet understood or need a better description, and to which extent all the models could be composed to a universal model of metal halide lamps.

3. MODELLING OF THE LAMP STATE

Besides of attaining a better understanding of the various phenomena in lamps the aim of modelling is to predict the electrical and radiative lamp data. The first model calculations that have been carried out for metal halide lamps considered only the arc column, which can be regarded as cylindrically symmetric for vertically burning lamps [1-3]. To obtain the desired results the energy balance of the plasma has to be solved together with the set of thermodynamic equilibrium relations that determine the local plasma composition. From these data the local transport coefficients, i.e. thermal and electrical conductivity as well as the net emission coefficients can be calculated and with these data the energy balance is solved.

Fortunately the poor knowledge of various input data for some of the metal halides used in lamps does not influence the results too much, as in all metal halide lamps mercury is the overwhelming species (typically present with a partial pressure of 3 to 30 bar). Partial pressures of metal halides usually are below 100 mbar. Thus radiative energy transfer in optically thick mercury lines is the dominant loss mechanism that determines the arc temperature profile. An exact knowledge of the broadening mechanisms that influence the shape of the mercury lines is indispensible for all lamp models, however. The quality of models for metal halide lamps largely depends on their ability to describe this energy transport accurately. So all the models developed for mercury lamps [4,5] must be regarded as the basis of all work on metal halide lamps.

To include the modifications due to the metal halide additives we must know their concentration in the gas phase. In all metal halide lamps the filling charge is much larger than the amount that can be vapourized in the burning lamp. This is to compensate for losses of metals due to diffusion or reactions with the wall material during the life of a lamp. So the first step to determine the plasma composition is to calculate the chemical equilibrium between the gas phase and the salt melt at the coldest spot of the inner boundary of the arc tube and from this the local composition all over the arc column. Very effective numerical programs have been developed in the last years to determine chemical equilibrium compositions in complex systems [6].

The number of components that may be present in a typical metal halide plasma may be very large. In a lamp containing the iodides of sodium, thallium and indium for example, the number of gaseous compounds that has to be considered is 56 besides of atomic mercury and the starting gas, even if we neglect the water and carbon impurities that are present in every metal halide lamp. Fig. 1 shows a plot of the partial pressures of the twelve most abundant species for the temperature range covered by the arc plasma.

In addition ,25 liquid and 69 solid compounds are known that might occur, if we include quartz and tungsten to the thermodynamic system.

The numbers given above do not even include mixed phases that can be formed in the salt melt. Such mixed phases may considerably influence the gas phase pressure of the halides. The mixture mentioned above is used in lamps with a content of 85% NaI. This leads to a reduction of the concentration of thallium in the gas phase by a factor of about ten, compared to the pure TlI system. Chemical modelling of such multicomponent salt melts is still at the beginning. First results from a generally applicable model have been recently presented [7,8].

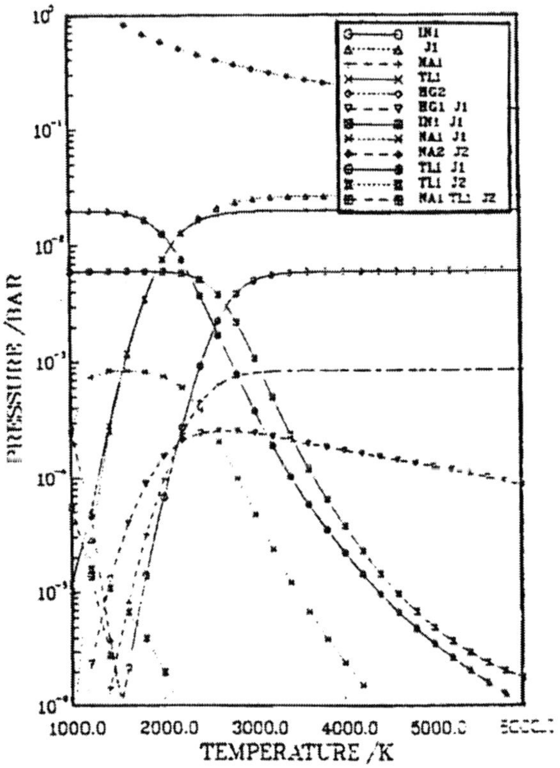

Fig.1 Partial pressures of the twelve most abundant species besides mercury in a lamp filled with iodides of sodium, thallium and indium.

But even if the conditions at the cold spot are perfectly known, this is far from beeing sufficient to calculate the local plasma composition exactly. Due to the combined action of diffusion processes and natural convection inside the lamp the components of the arc gas tend to segregate from each other [9-11]. The metal halides gather in the lower part of the lamp. In long verticaly burning lamps their concentration may vary by more than one order of magnitude from top to bottom. The segregation can be understood from the fact that the molecules that are present in the outer cold regions of the arc tube, in which the convective flow is directed downward, diffuse less quickly to the central region than the atoms do in the opposite direction. This leads to a net downward flux of the additives that must be compensated by a vertical concentration gradient. The strongest segregation effects are observed, when ambipolar diffusion of ions and electrons against a flux of neutral atoms is the driving force [10].

So any model of metal halide lamps must by principle be at least two-dimensional, even if the geometry is strictly cylindrical and end effects are neglected. Predicting segregation effects requires the combined solution of the energy balance and of the momentum balance to calculate the flow field. Numerical results for Na-Sc-lamps derived from a model on this level have been presented by DAKIN and SHYY [12]. In this model the influence of the electrodes is included in form of rough estimations of the voltage drop and energy loss at the electrode surfaces, while much numerical effort is given to an exact determination of the radiative processes in the arc column. But even in this model some simplifications had to be introduced to keep the computation time in reasonable limits.

The simplifications as made by DAKIN appear reasonable for long arc tubes, as they normally are used for high power lamps. In low power lamps, however, the tube diameter is of the same magnitude as the total arc lentgh. Models for such lamps must therefore include a more detailed description of the plasma in the electrode region, as the hot plasma spots in front of the electrodes contribute considerably to the total radiated power. Such models have by now been developed only with strongly simplified description of the radiation transport [13], so that they are suitable only to determine the temperature profile, but not to describe the radiative properties. Fig.2 shows the comparison of a measured temperature profile of such a lamp with a profile calculated by one of the authors. An extension of the numerical method to a more detailed description of metal halide lamps appears principically possible, but requires an enormous numerical effort, as the radiation transport has to be treated in a fully two- or even three-dimensional way.

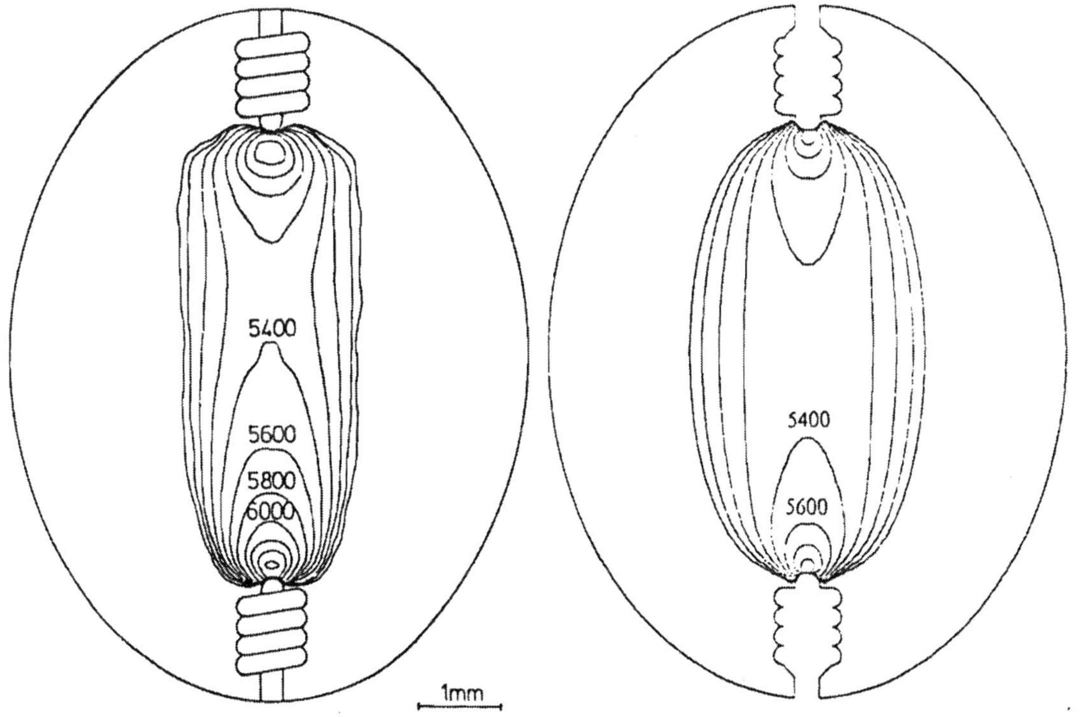

Fig.2. Measured and calculated temperature profiles of a 35W metal halide lamp containing NaI, TlI and InI.

A realistic view of the processes in the electrode boundary layers must also include non-equilibrium processes that determine the size and temperature of the plasma spots. These phenomena are strongly influenced by the electrical conductivity in the boundary layer and thus by the concentration of the additives. To calculate these concentrations we need not only a detailed picture of the convection and diffusion processes near the surface, but also the electrical field and space charge distribution. Today we are far away from having suitable numerical models for the electrode regions with an accuracy comparable to those for the arc column. A number of attempts has been made to describe at least qualitatively the DC behaviour of electrodes in high pressure discharge lamps with their transitions between diffuse and constricted spot mode [13-15], but there is still no final agreement on the question, which are the dominant processes that cause the formation of spots. In this point we must first put in order the physical models before we can hope to set up some realistic engineering tool.

For electrodes in metal halide lamps an additional effect must be considered: the change of thermionic emmissivity by the presence of metal atoms on the electrode surface which are deposited there from the plasma as a mono-layer. This effect is well known from experiments[16] in lamps containig iodides of thorium or scandium, but a quantitative description is still missing. So modelling of electrode phenomena remains a wide field for the future.

4. TRANSPORT AND CORROSION MODELS

Modelling of lamps does of course not only mean to achieve a good description of the actual state of a burning lamp, but should also give information on the processes that change this state during the burning time and finally lead to failure of the lamp. Metal halide lamps, due to their highly reactive ingredients, show various transport and corrosion phenomena that pose severe limits on filling and operating conditions to reach acceptable life times.

Modelling processes that take place on the "chemical time scale" does not necessarily require a detailed knowledge of the local plasma parameters. Many transport and corrosion processes can be described at least qualitatively, if only the conditions at the solid boundaries are known at which they take place. Over the last years the so called "solubility model" [17,18] has proved very successful. It is based on the assumption that the diffusion coefficients of all species in a carrier gas (mercury) are of the same order, so that the net diffusion flux of some element is proportional to the gradient of its "solubility", the sum of partial pressures of all the components that contain the respective element:

$$\Sigma_i = \sum_j \beta_{ij} p_j.$$

(β_{ij} are the numbers of atoms of kind i in the component j)

As an example we will discuss tungsten transport in a metal halide lamp containing NaI, TlI and InI. The solubility of tungsten and its change under various chemical conditions is shown in fig.3 for the temperature range at which solid tungsten may be in contact with the reactive gas. The plot on the left hand side shows that the pure metal halide mixture would lead to a transport that is always directed from the electrode tip (3000K) to the wall (1100K). Addition of small amounts of oxygen and excess iodine changes the picture completely. Due to the formation of gaseous WO_2I_2 the solubility is drastically enhanced below 1800K, so that the direction of transport is reversed and no tungsten is deposited on the wall.

Hydrogen impurities will however destroy this favourable effect (see fig.3b) and istead lead to strong transport from the colder parts of the electrode (1600K) to the wall. This process sets strong limits on the impurity level in lamps with a chemical wall cleaning cycle. An improvement of the situation can be achieved with the addition of arsenic to the lamp gas (see fig.3c), but this requires special measures to avoid other unwanted effects [19].

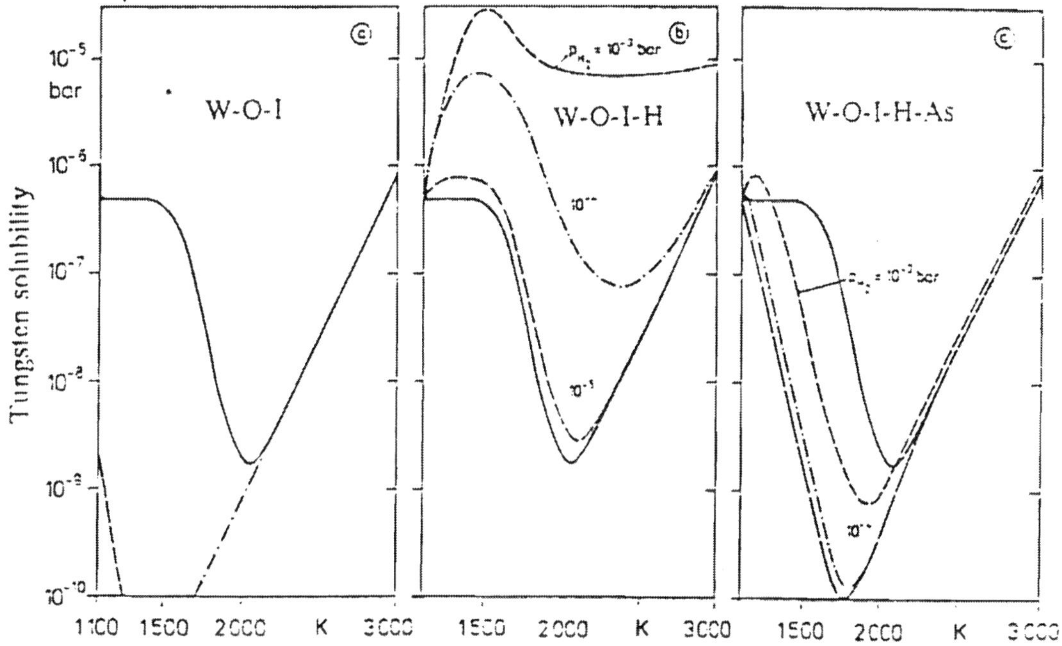

Fig.3. Tungsten solubility in metal halide lamps: a) ----- pure metal halide, ----- addition of oxygen, —— addition of iodine and oxygen. b) Influence of hydrogen impurities c) Influence of arsenic: —— without arsenic, — — with arsenic. -----, ----- with arsenic and hydrogen.

To obtain a quantitative picture about the decrease of light output of a lamp due to wall blackening we must of course go more into details. At the hot tip of the electrode the tungsten loss rate is not only a function of temperature. It also depends on the local electromagnetic field in the electrode boundary layer, as not only tungsten containing molecules and atoms contribute to the transport rate but also ions. We know that in many metal halide lamps tungsten is deposited at the arc root, where the temperature may be close to the melting point of tungsten. Modelling of these processes is possible only after a valid description of the stationary balance of the electrodes has become available.

But even if we have accurate information on the evaporation rate, to answer the question, where tungsten is deposited and in which form, requires the solution of the complete set of balance equations together with the local chemical equilibrium relations. Models as that of DAKIN [12] developed to describe stationary states of lamps can be extended in principle to yield also transport rates, regarding the transport processes as taking place in a sequence of stationary states with slightly changing chemical and physical boundary conditions.

It should be mentioned here, that models using thermodynamic equilibrium relations to determine the local flux density of the transported species may lead to severe errors in the determination of transport rates. As ROSNER [20] has pointed out, in many cases the concentration of the species in question may be so low that the collision rate is insufficient to establish local reaction equilibrium in the gas phase. Such phenomena should better be described by a "frozen chemistry" model, in which gas phase reactions are completely ommitted and equilibrium is established only at the solid boundaries.

While tungsten transport is a process in which only the lamp filling and the electrodes are involved, there are other long-term changes of the lamp properties that result from interactions between the metal halides and the quartz tube. Some of the reactions include the chemistry of tungsten as well. Common to all these processes is that they lead to an increase of the pressure of free iodine in the lamp, that means iodine not bound in metal halides at wall temperature.

The presence of free iodine does not only influence the arc voltage as it leads to a reduction of the electrical conductivity, it may even produce strong changes in the arc temperature profile so that the radiative properties of the lamp will vary with time. To predict such variations it is necessary to have appropriate models that describe the production of free iodine. Four causes are known from which the iodine can originate:

1. Water contained in the wall material leaks out during operation and reacts with the metal iodides forming oxides and free iodine, while the hydrogen leaves the lamp by diffusion through the quartz wall [21].

2. In lamps containing NaI sodium ions may diffuse through the quartz tube leaving the iodine back in the lamp [16]. Though this process is well known, quantitative modelling of the loss rate is not yet possible as it is still a controversy, to which extent free electrons on the outer side of the tube wall contribute to the driving force of the process [22].

3. Direct reactions are known to take place at high wall temperatures between quartz and the iodides of scandium and rare earth metals:

$$4ScI_3 + 7SiO_2 \rightarrow 2Sc_2Si_2O_7 + 3SiI_4$$

$$2LnI_3 + SiO_2 \rightarrow 2LnOI + SiI_4$$

(Ln stands for any metal of the lanthanide group). The SiI_4 decomposes at high temperatures. The silicon is dissolved in the electrode material, while free iodine remains in the gas phase [23,24].

4. Near the base points of the electrodes (temperature range 1400K -1500K) sodium iodide reacts with quartz and tungsten according to[21]

$$W + 2SiO_2 + 2NaI \rightarrow Na_2WO_4 + SiI_4$$

with the subsequent steps of reaction being the same as mentioned before.

Quantitative modelling of these reactions is not possible with thermodynamic equilibrium models. Applying such a model to the reaction between scandium iodide and quartz would yield the result that it is impossible to build a scandium lamp, as all the scandium iodide would be transformed to silicate immediately. Similar unrealistic results are obtained for the formation of sodium tungstate, which would not stop before the iodine pressure reaches a value of several bars. So for a quantitative description of free iodine formation we need models that are completely new for lamp researchers, models that include kinetic processes at the surface of the arc tube and of the electrodes.

5. FUTURE ASPECTS

In this paper we have tried to give a short overview on the topics that are relevant in modelling of metal halide lamps. Let us now strike the balance, what has been reached and which should be the main fields of interest for the future.

We have seen, that with the aid of powerful computers it is possible to treat such complex phenomena as they occur in metal halide lamps quantitatively. We can go more and more into the details of the balance equations and introduce all the couplings between the different parts of the lamp by iterative numerical methods. A lot of skilful work will be necessary to develop computer codes that give reliable results with acceptable effort. But we must keep in mind that there should always be somebody who is realy interested in the results we obtain.

It is questionable if it will ever be reasonable to design a metal halide lamp by computer modelling, as even with increasing computational power there remains the work of generating the computer code. Of course this question must be answered anew from time to time when more efficient software packages become available to solve some of the partial problems and when more complete data files exist on the properties of all the "exotic" elements that may be present in metal halide lamps.

But modelling of metal halide lamps does not necessarily mean simulating of lamp properties. There are also some aspects, where the physical understanding of the phenomena is still insufficient. Material transport near the electrode surfaces and corrosion processes are far from being understood quantitatively. To develop predictive models for these interactions that determine the long-term changes of the lamp state would be of mayor interest for all lamp manufacturers, as such model calculations could replace many long-term tests, which now with rated life times of 5000 to 10000 hours take years, before final results can be stated.

But such models are impossible without a detailed model description of the actual lamp state, as this model describes the surrounding, in which all the transport and reaction processes take place. So we should stay on the path of steadily improving modelling of the lamp state, but we should keep in mind that this effort will pay off only if we also succeed in quantitative modelling of the processes on the "chemical time scale".

REFERENCES:

[1] R.O.SCHAFFNER, Proc. IEEE 59, 622 (1971)
[2] E.FISCHER, J. Appl. Phys 45, 3365 (1974).
[3] J.T.DAKIN and R.P.GILLIARD, J. Appl. Phys. 62, 79 (1987).
[4] W.ELENBAAS, The High Pressure Mercury Vapour Discharge, (North Holland, Amsterdam, 1951).
[5] R.J.ZOLLWEG, J.J.LOWKE and R.W.LIEBERMANN, J. Appl. Phys 46, 3828 (1975).
[6] E.SCHNEDLER, Calphad 8, 265 (1984).
[7] H.GREINER, Mathl. Comput. Modelling, 10, 529 (1988).
[8] H.GREINER and E.SCHNEDLER, Proc. Symp. High Temp. Lamp Chem. II, p.111, (Atlanta, USA,1988).
[9] E.FISCHER, J. Appl. Phys 47, 2954 (1976).
[10] H.P.STORMBERG, J. Appl. Phys. 52, 3233 (1981).
[11] R.J.ZOLLWEG, R.W.LIEBERMANN and D.K.MC LAIN, J. Appl. Phys. 52, 3293 (1981).
[12] J.T.DAKIN and W.SHYY, Proc. Symp. High Temp. Lamp Chem. II, p.139, (Atlanta,USA,1988).
[13] E.FISCHER, 4th Int. Symp. on the Science and Techn. of Light Sources, Karlsruhe (1986) and Philips J. Res. 42, 58 (1987).
[14] J.F.WAYMOUTH, J. Light & Vis. Env. 6, 53 (1982).
[15] R.H.SPRINGER and W.H.LAKE, Journal of IES 13, 304 (1984).
[16] J.F.WAYMOUTH, Electric Discharge Lamps (M.I.T. Press, Cambridge, Mass., 1971).
[17] H.SCHÄFER, Z. anorg. allg. Chem. 400, 242 (1973)
[18] G.DITTMER, A.KLOPFER and J.SCHRÖDER, Philips Res. Repts. 32, 341 (1977).
[19] E.FISCHER, Proc. Symp. High Temp. Lamp Chem., p.47, (Toronto, Canada, 1985).
[20] D.E.ROSNER, Proc. Symp. High Temp. Lamp Chem. II, p.111, (Atlanta, USA,1988).
[21] E.FISCHER, Proc. Symp. High Temp. Lamp Chem. II, p.90, (Atlanta, USA, 1988).
[22] F.NAGEL and O.KAPOSI, Proc. Symp. High Temp. Lamp Chem., p.72, (Toronto,Canada,1985).
[23] D.M.SPEROS, R.K.DATTA and D.A.LYNCH, J. Electrochem. Soc. 128,778 (1981).
[24] W.VAN ERK and T.RIETVELD, Proc. Symp. High Temp. Lamp Chem., p.57, (Toronto, Canada, 1985).

EXPERIMENTAL INVESTIGATIONS OF Na REACTIONS IN HIGH PRESSURE SODIUM (HPS) LAMPS

R. K. Datta
Advanced Technology Department, General Electric Lighting
Nela Park, Cleveland, Ohio 44112 U.S.A.

In HPS lamps, elemental sodium vapor is used as arc material inside a hermetically sealed, dense, polycrystalline alumina (PCA) arc tube. A barium-calcium-aluminate is usually used as a sealant. During lamp operation, sodium reacts with Al_2O_3 (PCA), impurity oxygen, etc., resulting in sodium depletion leading to lamp-voltage rise and ultimate lamp failure. Several reactions for Na-fixation in HPS lamps have been postulated with their theoretical thermomechanical data:[1,2,3]

$2Na + 1.33\ Al_2O_3 \rightarrow 2Na\ AlO_2 + 2/3\ Al$ 1a

$2Na + 11.33\ Al_2O_3 \rightarrow Na_2O.11Al_2O_3(\beta\text{-}Al_2O_3) + 2/3\ Al$ 1b

$2Na + O + Al_2O_3 \rightarrow 2Na\ AlO_2$ 2a

$2Na + O + 11Al_2O_3 \rightarrow \beta\text{-}Al_2O_3$ 2b

$2Na\ AlO_2 + 10Al_2O_3 \rightarrow \beta\text{-}Al_2O_3$ 2c

In addition, alkaline-earth (Ca, Mg) impurity in the PCA arc tube will initiate another set of reactions:[4,5,6]

$2Na + O + 2.75(CaO.6Al_2O_3) \rightarrow \beta\text{-}Al_2O_3 + 2.75(CaO.2Al_2O_3)$ 3

Another reaction between sodium (ℓ,ν) and one of the aluminate phases, zeolitic $12CaO.7Al_2O_3$ crystallizing in the seal composite have also been identified:[7]

$Na(\ell,\nu) + 12CaO.7Al_2O_3\ (\text{white}) \rightarrow 12CaO.7Al_2O_3{:}Na\ (\text{green})$ 4

In the present investigation, following the sealed Ta-tube technique,[5] equilibrium thermochemical and kinetics data of these reactions are generated.

Based on 88 runs with varying reaction intervals (3-14 days), temperatures (600-1100°C), PNa (0.02-4 atm and PO_2 (10^{-40} atm.air), the equilibrium conditions, kinetics and activation energy for the reactions 1a, b and 2a, b were determined. It has been shown that for the reaction 1b, above 900°C and 0.5 atm PNa, there is a good agreement between the theoretical and the present experimental data. The deviations observed at lower temperatures and PNa may be attributed to the formation of metastable β-Al_2O_3 phases and sluggish reaction rates. Reactions 2a, b and c proceed rapidly, and depending on the temperature (900-1300°C), the reactions are complete within 70-6 hours. It is also shown that increasing PO_2 raises β-Al_2O_3 stability to higher temperature. At 0.1 atm PNa, with increase of PO_2 from 10^{-40} to 10^{-30} atm, the stability of β-Al_2O_3 increases from 875 to 1050°C. The reaction no. 3 at 900°C proceeds only when PO_2 is greater than 10^{-35} atm.[6] In the reaction no. 4, the amount of elemental sodium incorporated in the zeolitic $12CaO.7Al_2O_3$ lattice increases with temperature and PNa; from 1 wt.% at 900°C and 0.15 atm., the amount of incorporated sodium increases to about 3 wt.% at 1100°C and 4 atm. PNa.

REFERENCES

1. F.A.Elrefaie and W.W.Smeltzer, J.ELECTROCHEM.SOC., 128(7), 1443-1447 (1981)
2. G.Rog, S.Kozinski and A.Kozlowska-Rog, ELECTROCHIMICA ACTA, 28(1), 43-45 (1981)
3. K.L.Luthra, High Temperature Lamp Chemistry, Ed. E.G.Zubler, THE ELECTROCHEM.SOC.PROC., Vol 85-2, 156-171 (1985)
4. P.Hing, J.I.E.S., 10, 194 (1981)
5. R.K.Datta, High Temperature Lamp Chemistry, Ed. E.G.Zubler, THE ELECTROCHEMICAL SOC.PROC., Vol.85-2, 271-291 (1985)
6. J.D.Hodge, op.cit., 261-269 (1985)
7. R.K.Datta, AM.CER.SOC.BULL., 65(10), 1372 (1986)

THERMODYNAMIC CONSIDERATIONS ON TRANSPORT PROPERTIES OF HALOGEN-FILLED FLOOD-LIGHT LAMPS

G. Dittmer, U. Niemann
Philips GmbH Forschungslaboratorium Aachen
Weißhausstraße, D-5100 Aachen, F.R.G.

Linear flood-light lamps operating at higher voltages generally consist of a coiled tungsten coil with a number of supports enclosed by a silica envelope; they are filled with Ar or Kr together with additives of reactive gases such as CH_2Br_2. The effective gas phase composition and the transport level within the lamp are determined by condensation of tungsten halides and oxyhalides controlled by the lowest bulb temperature and the respective hydrogen content.

Main obstacles for achieving a regenerative halogen cycle are the relatively large amounts of gaseous impurities originating from the coil material and the high temperature gradients along the wall and the coil. Life-time of such lamps is limited by a complex interaction of different mechanisms. The prevailing processes are the development of hot spots, transports along supported sections of the coil and the attack of cold leads.

Transport effects may be discussed on the basis of summed tungsten pressures SPW derived from chemical equilibrium calculations. An SPW curve calculated for typical conditions of flood-light lamps is presented together with respective partial pressures as a function of filament temperatures in fig. 1. As the summed tungsten pressures in colder parts of the lamp considerably exceed values valid for hotter parts of the filament a regenerative transport cycle can be sustained during life-time. After incorporating diffusion effects local mass balances within such lamps may be evaluated; arrows in fig. 2 visualize direction and intensity of tungsten transports.

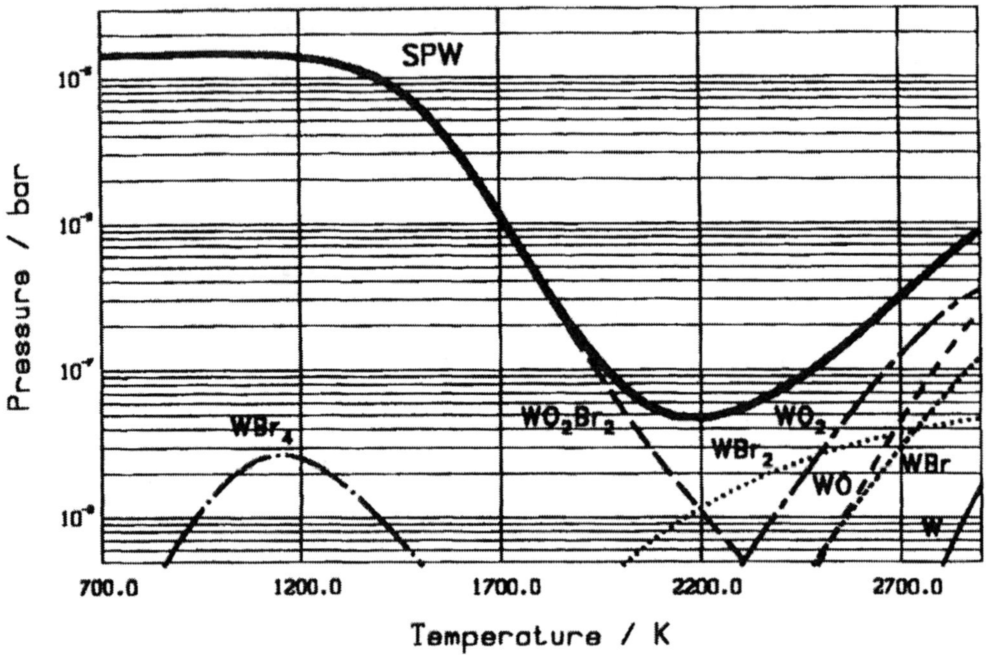

Fig. 1: Summed Tungsten Pressure and Partial Pressures in a Halogen Flood-Light Lamp with $p(CH_2Br_2) = 10^{-2}$ bar and $p(O_2) = 10^{-4}$ bar.

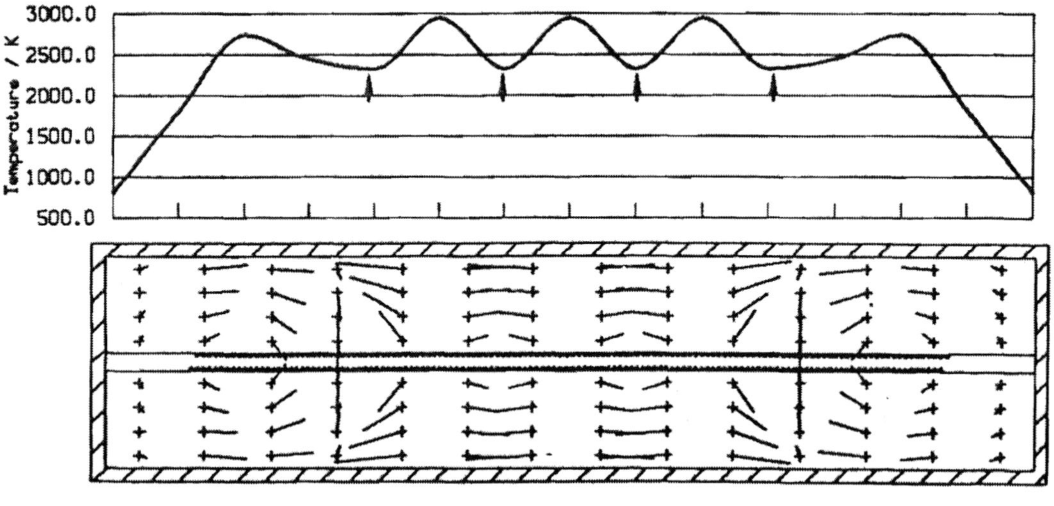

Fig. 2: Tungsten Transports in a Halogen-Filled Flood-Light Lamp with a Maximum Filament Temperature of 2900 K, Support Temperature of 2325 K, and Bulb Temperature of 700 K.

FLUID FLOW MODELLING OF A MODIFIED TRANSPIRATION EXPERIMENT DESIGNED TO INVESTIGATE ELECTRODE CORROSION

M.Duncan, J.Meads, and R.Devonshire
Chemistry Department, Sheffield University,
Sheffield S3 7HF, U.K.
N.W.O'Brien, S.A.Mucklejohn
Lamp Research, THORN Lighting Limited, Light Sources Division,
Melton Road, Leicester LE4 7PD, U.K.

Corrosion of the electrodes in metal halide discharge lamps is one of the major failure mechanisms governing their operating life. Since very little is known about the mechanisms of corrosion in these systems there is some interest in devising experiments which make the problem more accessible to quantitative study than it is in the lamps themselves. Experiments to investigate the reactions of tungsten with metal halides in the presence of controlled amounts of impurities under well defined conditions, and which were based on the transpiration technique, have been reported recently.[1] In the experiments a flow of inert gas (usually Ar with, or without, an accurately known concentration of an impurity such as O_2) passes over a liquid or solid metal halide sample at a rate slow enough to allow an equilibrium to be established between the condensed phase metal halide and its vapour. The equilibrated mixture then flows into a reactor cell which contains a tungsten filament operating at a known temperature. Tungsten loss is monitored continuously by observing changes in the filament's resistance and the overall loss obtained directly by weighing. Condensates derived from the reaction products appear on the walls of the reactor cell and in a collection tube downstream of the cell.

Fluid flow modelling of the gas/vapour transport in the modified transpiration system has been carried out to help understand the rôle of bulk, forced and convective, flows in the surface and bulk reactions between tungsten and species present in the gas in the reaction cell.[2] The modelling studies were also carried out to address the question of the extent to which kinetic effects contribute to the results obtained or, alternatively, to what extent such effects could be exposed by modifications to the operating conditions of the transpiration system.

The modelling proved of value in a number of respects[3]:

i) It showed the dominance of the hot filament in determining the reactor cell fluid dynamics for the conditions used in Ref.1;

ii) It was able to account qualitatively for the striking patterns of reaction product deposition that were observed experimentally;

iii) It identified possible new approaches to the recovery of quantitative information on the kinetics and thermodynamics of high temperature reactions which are based on the temperature-time histories that can be generated using a "particle" tracking facility;

iv) It led to the design of a novel "single-pass", vertically oriented, transpiration experiment;

v) It made possible calculations of the spatial distribution in the gas phase of the reaction between tungsten and oxygen.

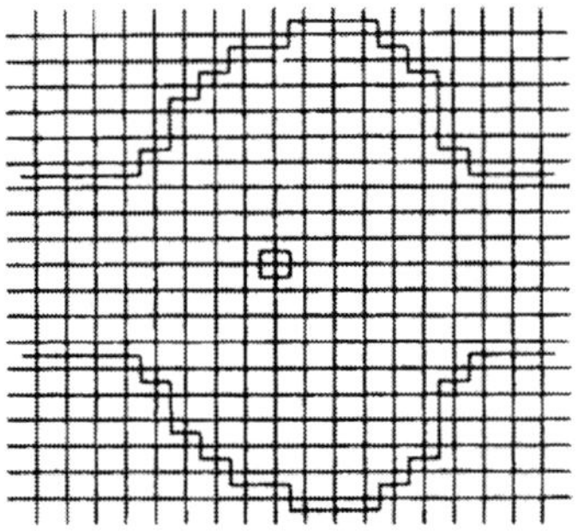

Figure 1a

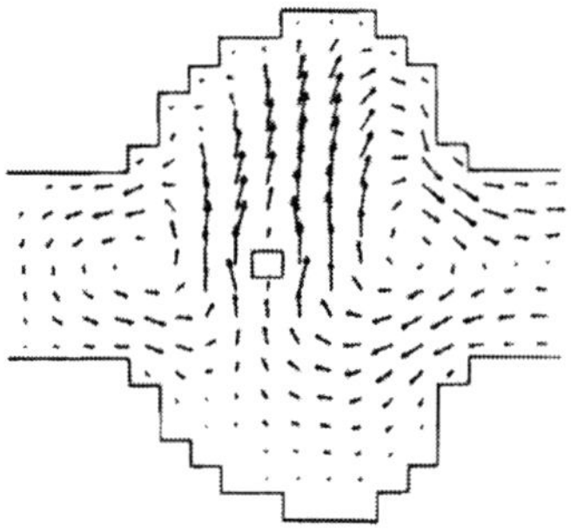

Figure 1b

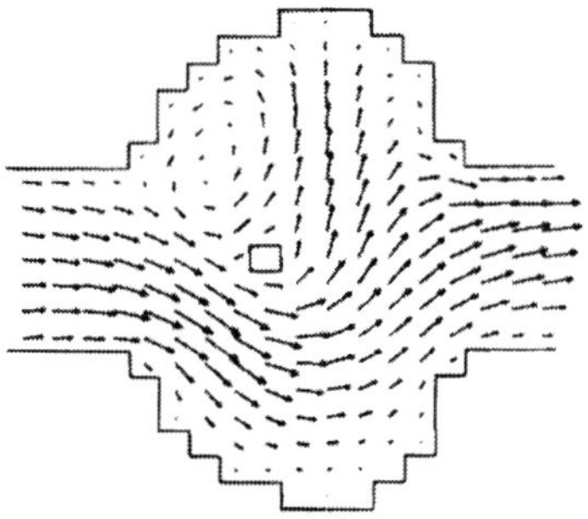

Figure 1c

Figure 1a shows the computational grid in the vertical symmetry plane passing through the reaction cell of the transpiration system. Figure 1b shows the velocity vectors in this plane for the conditions used in Ref.1 (T_{fil} 1750 K; T_{wall} 630 K). Figure 1c shows the effect on these vectors when the inlet gas velocity is increased by a factor of 10.

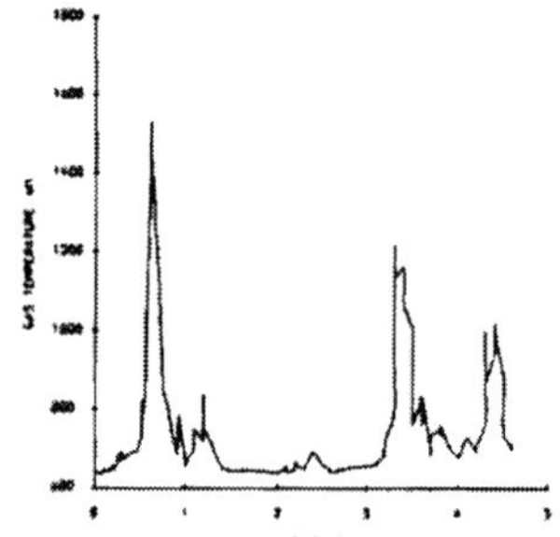

Figure 2a

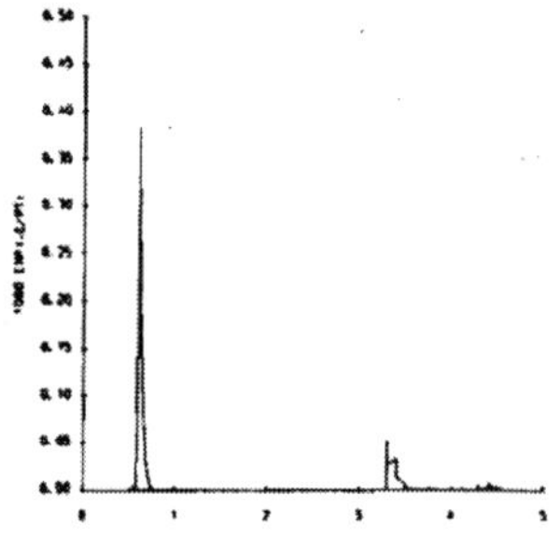

Figure 2b

Figure 2a shows the temperature-time history of a particle injected at a particular position in the inlet tube; Figure 2b the "reaction"-time history of the same particle assuming an activation energy of 100 kJ mol^{-1}.

References

[1] N.W.O'Brien, in "High Temperature Lamp Chemistry II," The Electrochemical Society, Proceedings Volume 88-4 (1988), 22-35.

[2] We used the finite difference code FLUENT (Creare R&D Inc.).

[3] M.Duncan, J.Meads, R.Devonshire, N.W.O'Brien, and S.A.Mucklejohn, to be published.

A TECHNIQUE FOR THE STUDY OF THE INTERACTIONS OF METAL HALIDES WITH GASEOUS IMPURITIES

M.Fields, S.A.Mucklejohn and N.W.O'Brien

Lamp Research, THORN Lighting Limited, Light Sources Division, Melton Road, Leicester LE4 7PD, U.K.

The introduction of certain impurities into metal halide discharge lamps during manufacture gives rise to phenomena which severely limit performance and life of the product.[1] Oxygen and water are particularly detrimental since they promote excessive transport of tungsten in the lamp. Whilst studying the corrosion of hot tungsten in the presence of metal halide vapours and oxygen,[2] it became apparent that it was essential to understand fully the reaction of the condensed phases of metal halides and oxygen over a wide range of temperatures and concentrations.

The experimental apparatus and technique are based on the transpiration system described in detail in reference [3]. This work was first reported at the 1988 EUCHEM Conference on Molten Salts.[4] Argon is used as the carrier gas and oxygen is introduced into the gas stream from a certified (argon + oxygen) supply. All gas flows are regulated by thermal mass flow controllers. Typically the total gas flow is 25 cm^3 per minute and the oxygen concentration is 100 vpm. The oxygen concentration in the emergent gas stream is continuously monitored using a fuel cell detector. The upper temperature limit for these experiments is governed by the range of temperatures over which the transpiration technique is valid; i.e., the maximum metal halide pressure is approximately 11 kPa.

Results are presented for the interaction of oxygen with:

$SnCl_2$(s,l); SnI_2(s,l); NaI(s,l); InBr(s,l); InI(s,l). Experimental studies have been complemented by chemical composition calculations. The reaction scheme for $SnCl_2$+ O_2 is given below.

$$SnCl_2(l) + O_2(g) = SnO_2(s) + Cl_2(g) \quad (1)$$

$$Cl_2(g) + SnCl_2(g) = SnCl_4(g) \quad (2)$$

Reaction (1) is not quantitative under these conditions and some oxygen remains in the gas stream. Similarly, reaction (2) does not consume all of the available chlorine. A detailed account of this work is included in reference [2].

The technique outlined here can readily be adapted to study the kinetics and reactions of metal halides with other gaseous species. The technique benefits from the fact that conditions can be accurately and reproducibly controlled over a wide range of additive concentrations. This approach also enables the effect of one additive to be studied without interference from others. Thus, the effect of oxygen on metal halides can be quantified without interference from water vapour and vice versa.

References

[1] S.A.Mucklejohn, Proc. Symp. High Temp. Lamp Chem., editor E.G.Zubler, The Electrochemical Society (1988).

[2] N.W.O'Brien, Proc. Symp. High Temp. Lamp Chem., editor E.G.Zubler, The Electrochemical Society (1988).

[3] N.W.O'Brien and S.A.Mucklejohn, J. Chem. Thermodynamics, 19, 1065 (1987).

[4] M.Fields, S.A.Mucklejohn and N.W.O'Brien, EUCHEM Conference on Molten Salts, St.Andrews (1988).

LASER-RAMAN SPECTROSCOPY OF METAL HALIDE VAPOURS AT HIGH TEMPERATURE

M.Fields and R.Devonshire
Chemistry Department, Sheffield University,
Sheffield S3 7HF, U.K.
H.G.M.Edwards and V.Fawcett
School of Chemistry and Chemical Technology, Bradford University,
Bradford BD7 1DP, U.K.

This work was undertaken in order to extract thermodynamic data (e.g., Gibbs free energies) for a number of high-purity metal halides used as additives in discharge lamps. Much of the available data for metal halide vapours were obtained at relatively low temperatures, and have to be extrapolated to the high temperatures found within the discharge lamp environment. Such extrapolations are unreliable, and require gross assumptions to be made about the system (e.g., chemical integrity). Spontaneous Raman scattering has been exploited in experiments designed to determine the required thermodynamic data at temperatures approaching those of the cool spot temperature of an operating lamp.

Preliminary investigations have involved the study of tin(II) chloride, thallium(I) bromide and indium(I) iodide as single component systems. The samples were sealed in silica ampoules and heated in a resistance wire furnace designed for optical studies up to 1100 K. The 488.0 nm argon-ion laser line was used for excitation, and strong Raman signals were observed for each of the three vapours. Results obtained from the tin(II) chloride system at 701 K and 923 K are shown in Figure 1a and b respectively. Bands attributable to the monomer of tin(II) chloride are clearly visible at 352 cm^{-1} (v_1) and 120 cm^{-1} (v_2), in good agreement with previous work.[1] The broader features in the spectrum are attributed to a dimeric tin(II) chloride species. Force constant calculations have been carried out to verify the assignments. The bands attributable to the dimer species show a marked decrease in intensity (relative to the monomer bands) with increasing temperature.

The asymmetric stretch of the monomer (v_3), is an inherently weak signal in the gas phase, and is partly obscured by the stronger v_1 band. Analysis of the spectrum to deconvolute this feature (isolated at 334 cm^{-1} by a previous matrix-infrared study[2]) is required prior to calculation of the thermodynamic quantities. Two methods are then available for the calculation of thermodynamic quantities from the Raman data[3]:

i) direct comparison of accurately measured line intensities, as a function of temperature, for each of the species present, in order to calculate the mole fraction and its derivative with respect to temperature. (A corrected form of the line intensity is proportional to mole fraction).

ii) determination of the Raman scattering cross sections of each species in order to calculate the mole fractions.

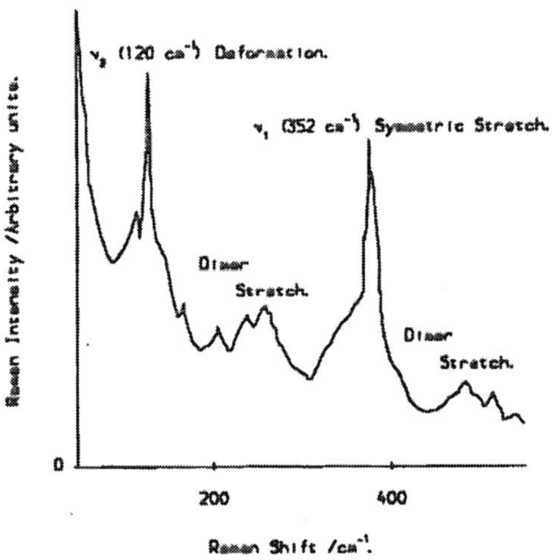

Figure 1a

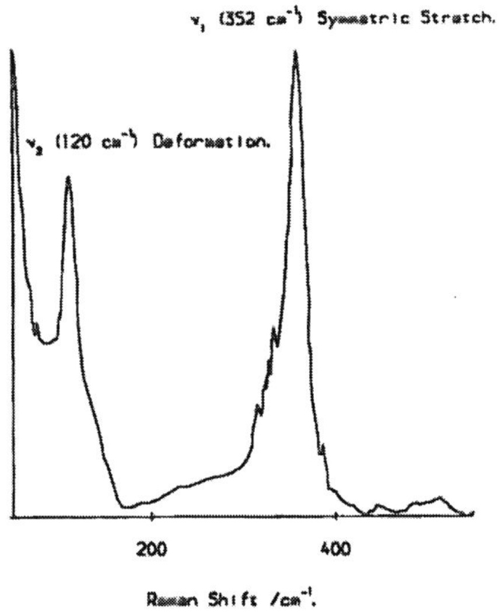

Figure 1b

Our observations are in qualitative agreement with a previous mass spectrometric study [4] of the vapour phase equilibria above tin(II) chloride over the temperature range 576 K to 694 K.

With both the thallium and indium systems, strong resonance fluorescence progressions (with at least twelve bands) have been observed. The Raman signals are partially obscured by the fluorescence, and methods to suppress/avoid these background signals are currently under investigation.

A preliminary account of this work has been given [5]; a full account of the work is to be published.[6]

Acknowledgements.

The authors gratefully acknowledge the help of Dr. S.A. Mucklejohn and Mr. N.W. O'Brien (THORN Lighting Limited (T.L.L.)) with sample preparation, and Mr. D.W. Farwell (B.U.) for technical assistance. M.F. acknowledges financial support from S.E.R.C. and T.L.L.

1. R.O. Perry, *J.C.S., Chem. Comm.*, 886 (1969) and I.R. Beattie, R.O. Perry, *J.C.S. (A)*, 2429 (1970).

2. L. Andrews, D.L. Frederick, *J.A.C.S.*, 92(4), 775 (1970).

3. P. Lenain, E. Picquenard, J. Corset, D. Jensen, R. Steudel, *Ber. Bunsenges. Phys. Chem.*, 92, 859-70 (1988).

4. N.W. O'Brien, S.A. Mucklejohn, *J. Chem. Thermodynamics*, 19, 1065 (1987).

5. M. Fields, R. Devonshire, H.G.M. Edwards, V. Fawcett, Molten Salts Discussion Group Summer Meeting, Leeds, July 1989.

6. M. Fields, R. Devonshire, H.G.M. Edwards, V. Fawcett, (in preparation.)

THERMOCHEMICAL DATA OF SCANDIUM HALIDES EVALUATED FROM VAPOUR PRESSURE MEASUREMENTS AND REACTIONS WITH QUARTZ

E. Fischer, R. Kaiser, U. Niemann
Philips GmbH Forschungslaboratorium Aachen
Weißhausstraße, D-5100 Aachen, F.R.G.

Total vapour pressures above ScX_3 compounds (X = I, Br, Cl) have been measured in the temperature range relevant for high pressure discharge lamps (900-1200 K) using a quartz spiral pressure gauge (for details see (1)). Equilibria with quartz were studied in order to determine pressures of gaseous reaction products (e.g. SiI_4) and to establish the constitution and composition of solid reaction products (silicates and/or oxyhalides). The latter information was obtained from a combination of X-ray measurements with experiments realizing a total evaporation of the respective scandium halide within the temperature range of measurement.

In the ScI_3/SiO_2 system only $Sc_2Si_2O_7$ was isolated; thermochemical data for this compound were improved via calorimetric measurements. Corresponding experiments on the $ScCl_3$ and $ScBr_3$ systems proved that additionally oxyhalide phases were formed: In the case of $ScCl_3$ final pressure measurements indicated the composition Sc_3O_4Cl whereas for $ScBr_3$ the composition (presumably ScOBr from X-ray pattern) could not be identified unequivocally by this procedure.

These informations were combined to evaluate a complete set of thermodynamic data of scandium halides. Partial pressures above ScI_3/SiO_2 calculated applying these data (ratio of gaseous ScI_3/Sc_2I_6 from (2)) are presented in fig. 1. The linear relationship observed on plotting atomic enthalpies of formation of ScX_3 compounds as a function of electronegativities of halogens (fig. 2) confirms the consistency of derived data (for details concerning this method see (3)).

References:

1. E. Fischer, R. Kaiser, Proc. Sympos. Sci. Tech. High Temp. Light Sources, Toronto, Canada, 1985, Proc. Vol. 85-2, 218 (1985)
2. J.H. Dettingmeijer, H.R. Dielis, B.J. de Maagt, P.A.M. Vermeulen, J. Less-Common Met. 107, 11 (1985)
3. G. Dittmer, U. Niemann, Philips J. Res. 42, 15 (1987)

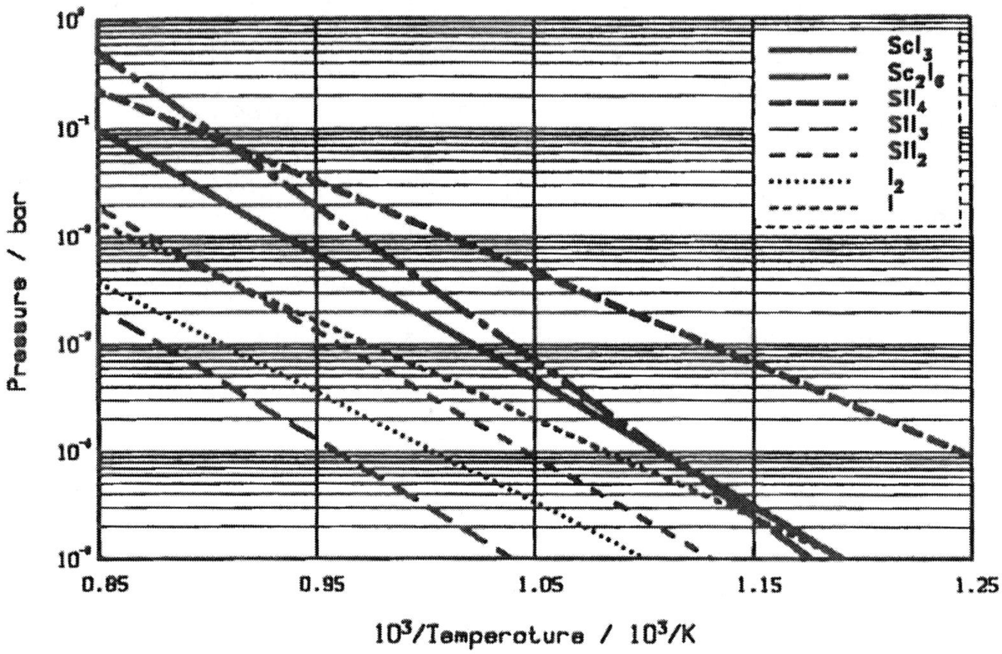

Fig. 1: Partial Pressure Composition above the Reaction System ScI_3/SiO_2

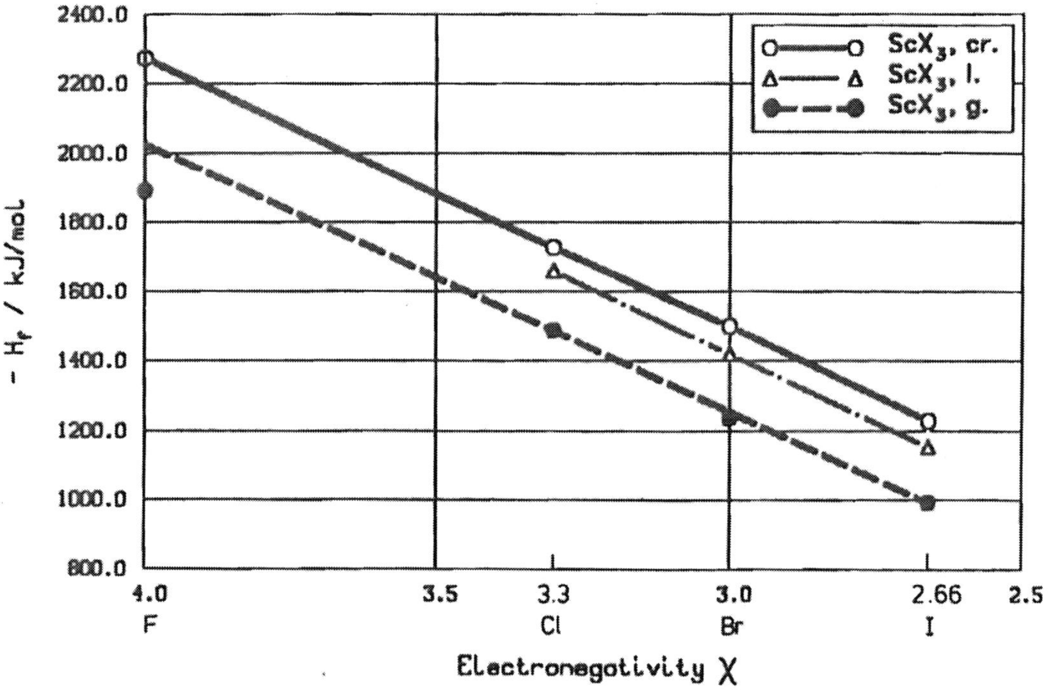

Fig. 2: Atomic Enthalpies of Formation of Solid, Liquid and Gaseous Scandium Halides as a Function of Electronegativity of Halogens

CHEMICAL VAPOUR TRANSPORT AND COMPLEXATION IN THE NaI-ScI_3 SYSTEMS

K. Hilpert, M. Miller[a], and M. Sai-Baba[b]
Institute for Reactor Materials,
Nuclear Research Centre (KFA) Jülich,
D-5170 Jülich, Federal Republic of Germany

a) On leave from: Technical University of Wroclaw, Wroclaw, Poland
b) On leave from: Indira Gandhi Centre for Atomic Research, Kalpakkam, India

Metal halide lamps containing mixtures of NaI and ScI_3 are of large economic importance. For the modelling of such lamps it is necessary to know the thermochemistry of the vapours of the NaI-ScI_3 system. The determination of the sodium and scandium enhancement by the formation of gaseous hetero complexes is of particularly large practical importance in addition to the thermochemical data for the dissociation of these complexes. This was the aim of our investigations carried out by Knudsen effusion mass spectrometry, KEMS, with a two compartment Knudsen cell and by the use of the vapour isolation method, VIM, developed by us. The use of a two compartment Knudsen cell for the first time rendered possible the elucidation of the fragmentation of gaseous species in the NaI-ScI_3 system, which is a prerequesite for the determination of reliable data by KEMS. The work on the NaI-ScI_3 system is part of our systematic investigations on the chemistry of metal halide lamps (see Ref. (1) and references quoted therein).

VIM allows the direct determination of enhancement factors at high temperatures. However, the integral data determined by this method cannot be explained without the information on the vapour composition obtained by KEMS. On the other hand, the determination of enhancement factors at the temperatures present in metal halide lamps is difficult by KEMS due to the demand for molecular flow in the effusion orifice of the Knudsen cell. The two methods, therefore, complement each other.

The gaseous species NaI, $(NaI)_2$, ScI_3, $(ScI_3)_2$, $NaScI_4$, and Na_2ScI_5 were identified in the equilibrium vapour of the $NaI-ScI_3$ system. The following thermodynamic data result for the reactions

$$NaScI_4(g) \rightleftharpoons NaI(g) + ScI_3(g) \quad \text{and} \qquad (1)$$

$$Na_2ScI_5(g) \rightleftharpoons 2NaI(g) + ScI_3(g) \qquad (2)$$

by KEMS:

$\Delta H^\circ_{298}(1) = 226.3 \pm 6.6$ kJ mol^{-1}, $\Delta H^\circ_{298}(2) = 388.1 \pm 8.7$ kJ mol^{-1}, $\Delta S^\circ_{298}(1) = 162.7 \pm 10.6$ J mol^{-1} K^{-1}, and $\Delta S^\circ_{298}(2) = 310.1 \pm 12.8$ J mol^{-1} K^{-1}.

The following enhancement factors E result by VIM at 1078 K for Na and Sc for the different condensed phase compositions given in parentheses: E(Na) = 3.0, E(Sc) = 0.29 ($x_{NaI} = 0.82$, $x_{ScI_3} = 0.18$) and E(Na) = 1.4, E(Sc) = 0.05 ($x_{NaI} = 0.97$, $x_{ScI_3} = 0.03$). By assuming ideal behaviour the equilibrium constant at 1078 K for reaction Eq. (1) was evaluated from the data obtained by VIM as 1.89×10^{-3}. A value of 1.65×10^{-3} was obtained by KEMS at the same temperature. The good agreement demonstrates the compatability of the data obtained by the two different methods.

The Na enhancement factors determined for the $NaI-ScI_3$ system are compared with those obtained by us for the $NaI-SnI_2$ and $NaI-PbI_2$ systems using VIM.

1. K. Hilpert, J. Electrochem. Soc., in press.

Immunization of quartz surface againts the crystallization and diffusion

S. Jonas, A. Kwatera, W. Ptak, E. Walasek
Institute of Materials Engineering, Cracow, Poland
B. Lesiuk, W. Stanioch
Research Development Center POLAM, Warsaw, Poland

A goal of the experiment was the increasing the immunization of the quartz surface to be used to halide lamps against aggressive action of the discharge medium.

Silicon nitride was selected for the experiments taking into account his well-known high - temperature immunization against the chemical corrosion in the medium of the metal vapours.

The following requirements on silicon nitride layers were decided: (a) homogeneity and continuity (b) maximal transmittance in visible range of radiation (c) high resistance against thermal shocks (d) immunization against vapour of alkaline and rare earths.

Silicon nitride layers have been obtained by means of the method of precipitation from vapour phase.

The paper contains the description of the method of receipt of the amorphous layers of silicon nitride about the thickness equal to 0,2 - 0,6 μm to be made on the quartz glass ground as well as the experiments results: (1) phase composition of the silicon nitride layers (2) chemical composition on the boundary of layer and quartz ground (3) transmittance in the range of 250 - 800 nm (4) immunization against crystallization (5) life of halide lamps made with the use of silicon nitride layers. The application of such layers in the metal-halide lamps 400 W with dysprosium iodide have resulted decreasing the rate of the luminous flux fall - 10 % of difference after 1000 hours - as well as the decreasing the lamp damages caused by the recrystallization process.

It was concluded that the obtained result is the effect of increased resistance from the recrystallization of the bulb covered with silicon nitride layer in the discharge medium containing among others dysprosium iodide as well as the effect of the diffusion barrier generated by the layer limiting the impurities resulted from the quartz.

Measurements of Sodium Loss From Fused Silica Metal Halide Arc Tubes

William M. Keeffe and Vincent D. Meyer
GTE Products Corp
Danvers, MA 01923

Measurements of the loss of sodium through the fused silica arc tube of an operating metal halide lamp have been made. The experimental arrangement is diagramed in Figure 1. A mass analyzer adjusted to mass 23 is used to detect sodium ions and neutrals under various operating conditions. Ions only can be measured by turning off the filament in the ionization chamber of the mass analyzer. The output of the mass analyzer can be input to an oscilloscope to record the 60 Hz waveform of the ions and/or neutrals selected by the mass analyzer. This is synchronized with the waveform of the voltage applied to the top electrode of the lamp. The time-averaged change in the sodium loss is measured by varying the voltage labeled bias in Figure 1. This bias voltage controls the DC voltage difference between the arc tube and the mass analyzer.

The time-averaged rate of sodium loss increases as the DC bias on the arc tube is made more positive. This result is consistent with electrolytic transport of sodium ions through the arc tube; however, both ions and neutrals are detected at mass 23. The measured ratio of ions at mass 23 to the total of ions and neutrals is 0.16, considerably different from the ratio of 0.99 measured by Nagel, Szanto, Adams and Peter[1]. A mechanism proposed by Waymouth[2] for the production of neutrals from the arc tube requires the generation of photoelectrons. The grounded metal vacuum chamber and the metal parts of the mass analyzer are possible sources of photoelectrons. Additional evidence for the presence of both ions and neutrals of sodium is found in the waveform of the signal from the mass analyzer with the filament on (ions and neutrals) and with the filament off (ions only), as shown in Figures 2 and 3. Note the rapid increase followed by a decrease in the sodium ion and neutral signal shortly after the AC voltage on top electrode turns positive in Figure 2. The initial rapid increase is attributed to the neutralization of sodium ions on the arc tube surface by photoelectrons . The decrease which follows is attributed to the depletion of photoelectrons. This is to be contrasted with the smaller signal and slower changing mass 23 signal when the filament in the ionization chamber is off (Figure 3) and only ions are detected.

The measurements are all consistent with a mechanism of sodium loss by electrolytic transport through the fused silica followed by neutralization of the positive sodium ions by photoelectrons released from metal parts in the vacuum system. In addition, there is also evidence for some extraction of positive sodium ions from the arc tube surface without neutralization, despite the stronger surface binding forces for the ions.

1. Nagel, F. , Szanto, J., Adam, J., and Peter, I., CIE Journal 6, 47 (1987)

2. Waymouth, J. F., "Electric Discharge Lamps", MIT Press, Cambridge, MA (1971) ; Waymouth, J. F., Koury, F., Gungle, W.C. and Peterson, C., Illum. Eng. 62, 214 (1971)

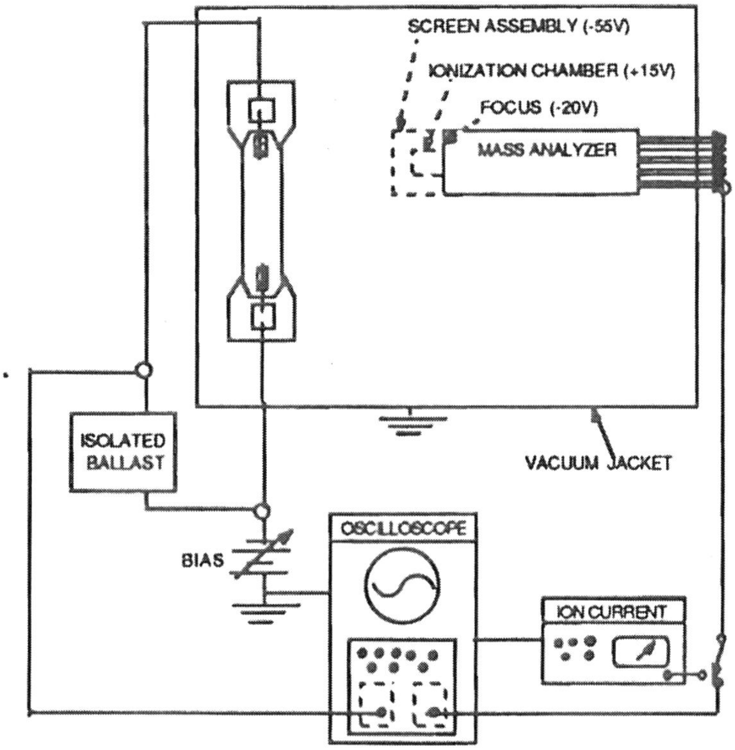

FIGURE 1 EXPERIMENTAL ARRANGEMENT

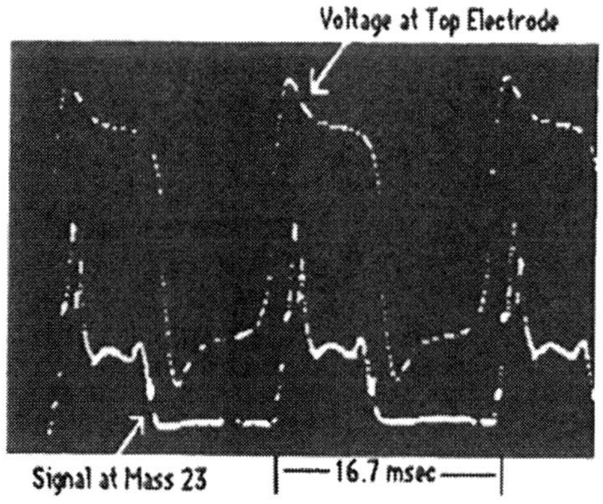

FIGURE 2 WAVEFORM FILAMENT 'ON'

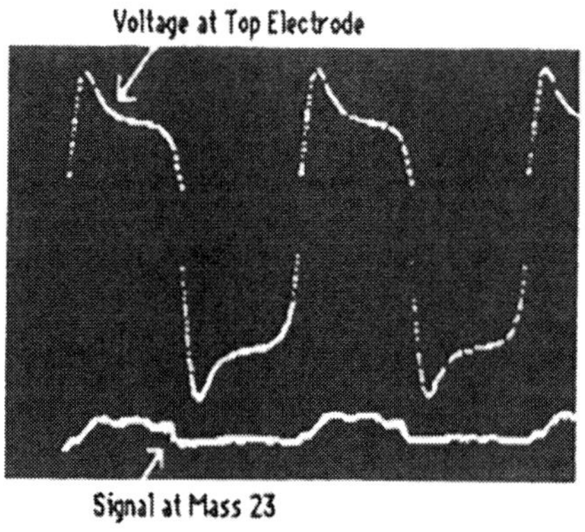

FIGURE 3 WAVEFORM FILAMENT 'OFF'

VAPOUR PRESSURE AND THERMOCHEMICAL PARAMETERS OF METAL HALIDES: InCl AND TlBr

S.A.Mucklejohn and N.W.O'Brien

Lamp Research, THORN Lighting Limited, Light Sources Division,

Melton Road, Leicester LE4 7PD, U.K.

T.R.Brumleve

Anderson Physics Laboratories, Inc.

406 North Busey Avenue, Urbana, Illinois 61801, U.S.A.

Continuing our investigations into the thermochemical properties of metal halide systems,[1] we have measured the vapour pressures and derived polynomials to describe the Gibbs free energies of formation of InCl and TlBr. We have reported comprehensive thermochemical parameters for InBr and InI recently.[2] A more detailed account of our work on the indium(I) halides will be published elsewhere.[3]

Vapour pressures were measured using the quasistatic technique over the ranges 528 to 914 K for InCl and 853 to 989 K for TlBr, the results can be represented by equations (1) and (2).

Vapour pressure above liquid InCl:

$$\ln(p/\mathrm{Pa}) = -10796(\mathrm{K}/T) + 23.309 \qquad (1)$$

Vapour pressure above liquid TlBr:

$$\ln(p/\mathrm{Pa}) = -12834(\mathrm{K}/T) + 23.392 \qquad (2)$$

Mass spectrometry shows that the vapours above both InCl and TlBr contain monomers and dimers. The mole fraction of $In_2Cl_2(g)$ is small and the analysis here is carried out assuming only the monomer to be present. Using values derived from a study of the (monomer+dimer) equilibrium, [4] the partial pressures of TlBr(g) and $Tl_2Br_2(g)$ can be represented by equations (3) and (4) respectively.

$$\ln(p/Pa) = -12479(K/T) + 22.988 \quad (3)$$

$$\ln(p/Pa) = -25153(K/T) + 32.971 \quad (4)$$

We have many reservations about the determination of (monomer+dimer) equilibria, especially at high temperatures [5] and are therefore studying the vapour phase equilibria above metal halides using the modified entrainment method.[6] Thermochemical parameters listed below for the TlBr species are derived assuming monomer only.

The temperature variation of Gibbs free energy change for the formation of InCl species from the gaseous elements can be represented by a polynomial, equation (5).

$$\Delta_f G_m(T)/kJ.mol^{-1} = A + BT + CT^2 + DT\ln(T/K) \quad (5)$$

	A	B/K	C/K^2	D/K
InCl(g)	-309.3871	-1.946E-2	-5.898E-7	7.986E-3
InCl(l)	-402.8717	1.466E-1	5.238E-6	-2.169E-3
InCl(s)	-435.0691	7.963E-1	1.596E-4	-1.087E-1
TlBr(g)	-237.7584	3.546E-2	8.715E-7	1.273E-4
TlBr(l)	-362.2507	4.115E-1	2.301E-5	-4.058E-2
TlBr(s)	-361.0097	1.915E-1	7.666E-6	-6.025E-3

References

[1] S.A.Mucklejohn and N.W.O'Brien, Lighting Research & Technology, 18, 88 (1986).

[2] T.R.Brumleve, S.A.Mucklejohn and N.W.O'Brien, Proc. Symp. High Temp. Lamp Chem., editor E.G.Zubler, The Electrochemical Society p.96 (1988).

[3] T.R.Brumleve, S.A.Mucklejohn and N.W.O'Brien submitted to J. Chem. Thermodynamics (1989).

[4] D.Cubicciotti, J. Phys. Chem., 68, 3835 (1964).

[5] N.W.O'Brien and S.A.Mucklejohn, J.Chem. Thermodynamics, 19, 1065 (1987).

[6] P.J.Gardner and P.Pang, Can. J. Chem., 66, 566 (1988).

Impurities in high pressure mercury discharge lamps

H.–P. Popp and R. Ascher
Lichttechnisches Institut, University of Karlsruhe, D–7500 Karlsruhe, FRG

High pressure mercury discharge lamps of similar power, dimensions and pressure but from various producers show differences in operating characteristics (e. g. ignition) as well as in radiation intensity. This oddity was supposed to be due to impurities of the vapour. Electronegative elements affect igniton, and elements with lower excitation potentials than mercury lower the plasma temperature. Regarding the manufacturing process lots of candidates for impurities are possible.

The spectra of several lamps were carefully recorded by a scanning monochromator of high resolution and a y–t–recorder. We confined the spectral range to the near UV. This region was scanned very slowly (0,5 nm per minute) and recorded with high sensitivity. The sensitivity range was determined by the very weak lines what of course involved the disadvantage that in the vicinity of the mercury lines the recorder was overdriven and so the lines appeared as broad plateaus which made it impossible to mark λ_0. The registrated spectra were very interesting: they showed a lot of lines! But the information could not be interpreted because of the uncertainty of the wavelength calibration.

There were two ways to overcome this problem:

- an insensitive spectrum was recorded (same scanning/recording speed, same spectral range) where the Hg line shapes could be seen and λ_0 be marked; the sensitive registration then could be interpreted by comparing with the insensitive one
- we designed a logarithmic amplifier to weaken the strong signals of the mercury lines

Both methods enabled us to determine the wavelengths of spectral lines and, by means of spectral tables, to attach them to elements. In dependance of the individual lamps Al, Fe, Mn, W, J, Zn, Cu ... were found as impurities (not all lines could be identified).

Summing up we can say that there are impurites in discharge lamps which can be made out by a very simple method. The knowledge about the impurities can lead to an improved – i. e.: purer – manufacturing process and to discharge lamps the operating characteristics and radiation intensities of which can be predicted with more precision.

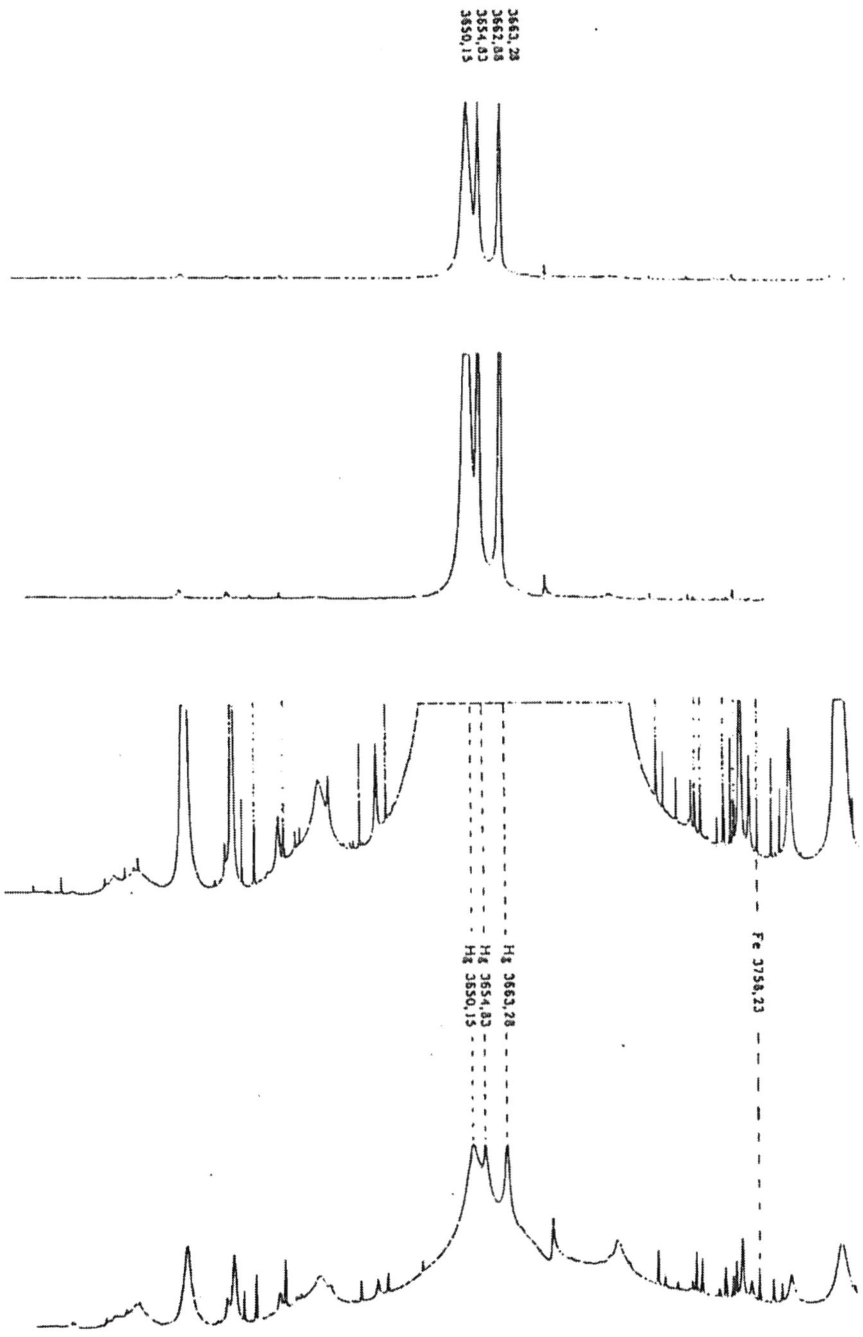

Gain in information:

a) "normal" spectrum (only the Hg–lines appear, their line shape can be completely made out (10 V/cm recorder sensitivity)

b) spectrum recorded with higher sensitivity, the Hg–lines are cut–off (5 V/cm recorder sensitivity)

c) again the sensitivity is increased, lines of impurities can be seen, the Hg–lines are a broad plateau (50 mV/cm recorder sensitivity)

d) use of a logarithmic amplifier: Hg–lines in completeness and lines of impurities appear (500 mV/cm recorder sensitivity)

CHEMICAL ANALYSES OF SODIUM, MERCURY, ALUMINUM AND TRACE ELEMENTS IN HIGH PRESSURE SODIUM LAMPS

G.D.A. Solleveld
Philips Lighting B.V.
Eindhoven
The Netherlands

In the inner envelope of a High Pressure Sodium (HPS) lamp amalgam (Na, Hg) and noble gas are present. Depending on the local temperatures during lamp operation (700-1300°C) Sodium can react with the PCA wall (Al_2O_3) (formation of some Sodium-Aluminum oxides) and with components of the frit, the emitter or the electrode (Mg, Ca, Ba, Sr, Y, Nb, W, etc.). To investigate and analyse these phenomena in more detail some analytical techniques have been developed.

1. Sublimation of the amalgam

During this procedure metallic Sodium and Mercury are removed from the discharge tube by heating at 800°C. First the sample discharge tube is opened in an Argon atmosphere and put in a bigger ceramic tube (one end closed), which is next gas tight closed with high vacuum seal. Then the end of the big tube, containing the sample, is heated to about 800°C, while another part is kept at low temperature by external cooling (cold Helium flow). In this way a cold spot is created where all the metallic Sodium and Mercury are collected. Next the

tensities of these continua depend on the alkali molar fraction, design of the burner and lamp load. They increase overall spectrum quality and may raise corresponding CRI.

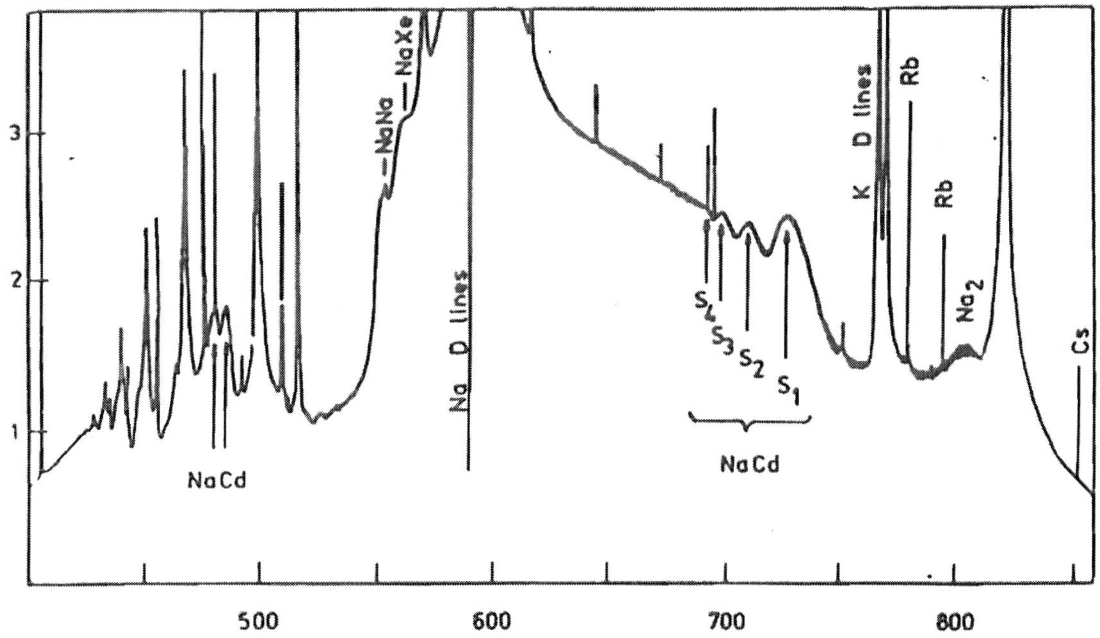

Fig.1. Spectrum of HP NaCd lamp at current maximum

Reported results contribute to the better understanding of the physical and chemical processes in HP alkali (and especially sodium) discharge. Furthermore, they indicate possible ways towards the improvements of HPS lamp CRI and spectrum quality. Finally, HP potassium/sodium lamps with proper mercury molar fraction are attractive for the optical pumping of the Nd:YAG lasers, and for some new medical applications (photodynamic therapy of skin cancer).

REFERENCES:

1. D.O. Wharmby, IEE PROC. 127 (1980) 165.
2. D. Veza, J. Rukavina, M. Movre, V. Vujnovic and G. Pichler, Opt.Commun. 34 (1980) 77.
3. J. Schleyen, J.J. Woerdman and G. Pichler, J.Mol.Spectry. 127 (1988) 1.
4. G. Pichler, D. Veza and D. Fijan, Opt.Commun. 67 (1988) 45.
5. G. Pichler, D. Fijan, D. Veza, J. Rukavina and J. Schleyen, Chem.Phys.Lett. 147 (1988) 497.
6. D. Fijan, D. Veza and G. Pichler, Chem.Phys.Lett. 154 (1989) 126.

HIGH TEMPERATURE SEAL OPERATION IN A MULTICOMPONENT HIGH PRESSURE METAL VAPOR LAMP

Richard Snellgrove & Elliot Wyner
GTE Products Corp
Danvers, MA 01923, U.S.A.

Color can be improved in high pressure sodium lamps by higher sodium pressure. This has led to commercial lamps with a CRI of 60 but with essentially the same 2100 K color temperature as standard HPS lamps. Further increase in sodium pressure raises color temperature, but heavily weights the spectrum to the red leading to color distortions. We have chosen to work with a four component system that provides radiation from sodium, thallium, cadmium, and mercury vapors, which yields a color temperature of about 2700 K, a CRI of 65-70, and a well balanced spectrum. Due to the low vapor pressure of thallium, cold spot temperatures must be elevated to 950°C - 1000°C, which is at minimum 150°C above present commercial HPS lamp operation. This accelerates several reactions not ordinarily a factor in HPS lamps to the point where they severely shorten the life of the multicomponent lamp.

The high temperatures cause early failure of the lamps from leakage of gas through the arc tube seals. Two distinct leakage pathways are observed, the inlead/frit interface and the alumina/frit interface, and in both cases the leakage can be attributed to reactions localized along that part of the seal. Each results in formation of channels which eventually breach the hermeticity of the arc tube.

Reduction of the oxide sealing material (frit) by the niobium of the inlead forms either gaseous species which diffuse away or solid phases which do not bond well to the inlead. Liquid fill moves into and expands the resultant gap, and thallium, the principal constituent of the liquid, then appears in the jacket and is found on the inlead surface after failure. With a Al_2O_3-CaO-BaO-MgO-B_2O_3 frit (F), niobium boride is the first phase formed and calcium niobate may develop later between the boride and niobium. For frits without boron, reduction and dissolution of alumina by niobium create the interfacial gap, and vapor may be leaked rather than liquid.

Both of the solutions to this problem that have been pursued eliminate contact of frit with the niobium inlead. One way to do this is to coat the niobium with another metal that is less reducing toward the frit. Molybdenum was used in coatings of several different thicknesses and successfully eliminated leakage along the inlead interface. No reaction could be detected at the interface of frit with molybdenum. Another successful strategy was to join alumina directly to the niobium inlead by sintering at 1900 °C. Reaction between alumina and niobium is not sufficient at the operating seal temperatures to open up a leakage pathway.

Reaction between alumina (PCA) and F frit produces a zone of high alumina content calcium aluminate phases adjacent to the PCA that increases in thickness with time and temperature of lamp operation. The reaction zone consists primarily of CA_2 sandwiched between borders of CA_6 and BC_2A_4. When the reaction zone reaches sufficient thickness, on the order of 10 μm, a system of cracks develops in the CA_2 which may eventually allow vapor to escape from the arc tube. The use of other frit compositions which contain little calcium and have higher melting points diminishes the reaction with PCA and the consequent cracking, at least that due to CA_2.

Chemical and Materials Problems Affecting Lamps

Dennis B. Shinn
GTE Electrical Products
Danvers, MA 01923

I. Introduction

The organizers of this Symposium suggested I review the relevant work on chemical and materials problems in lamps, including work in other laboratories, take a wide view of the past and point to future possibilities. Of course, coverage of all the current lamp materials problems is not possible. In this paper an attempt will be made to cover some significant issues not extensively addressed by others.

II. Tungsten-Halogen Lamps for Replacing Traditional Incandescents

Materials and chemical problems of tungsten-halogen lamps focus on the industry trend to low-wattage high-voltage sources and longer life. Tungsten filaments for low-wattage high-voltage lamps are very fine. For high efficiency and long life, the chemical environment must limit filament erosion to thermal evaporation only, while keeping the bulb clean. Consequently, tungsten-halogen lamps and materials have been studied extensively by high temperature modelers[1], materials scientists and experimentalists. Yet, lamp designs are still proliferating. Lamps for identical applications may use different envelope materials, halogen additives, getters, fill gas composition and coatings for seal protection or energy conservation. A large degree of Edisonion art remains in the tungsten-halogen lamp, notwithstanding an in depth scientific understanding.

Tungsten-halogen lamp replacements for traditional incandescent lamps offer the attractive features of general service lamps (instant light, low cost, dimmability, compact) plus higher efficacy. Double-enveloped lamps consisting of tungsten-halogen capsules inside outer bulbs are sold in PAR and A-line designs in the U.S. Typical lamps operate with filament temperatures above 2800 K for high efficiency and are filled with several atmospheres of rare-gas to retard filament evaporation. Either fused silica or hard glass capsules are used. Soft glass envelopes with treated inner surfaces have been described[2], but would be limited in temperature.

Even hard glasses are used near their temperature limit, about 550 °C for GE180 aluminosilicate glass. Halogen sources for maintaining the regeneration cycle include hydrogen halides and methyl halides. Kiesel[3] reported using CH_2ClI to limit carburization of the filament.

Consider the materials problems of an incandescent-replacement lamp of a hard glass W/HBr/Kr capsule contained in a soft glass outer bulb. Such lamps are available from 42 to 100 watts in the U.S. Capsules with wall loadings from 4 to 18 W/cm^2 operate at 10 to 15 atm pressure and 250 to 550 °C wall temperature. Tungsten primary filaments are 0.02 to 0.1 mm in diameter. Careful control of chemistry and design are necessary to keep evaporated tungsten off the walls without corroding the filament. As noted by Dittmer and Niemann[4], hydrogen pressure affects tungsten transport in the system $W/O_2/H_2/Br$. If significant H_2 is lost through the glass, corrosion of the cool portions of the filament can shorten life. The lower permeability of H_2 in hard glasses than in fused silica is advantageous in this lamp[5]. However, sodium from the glass can shorten life and increase wall blackening by the reaction with HBr to form NaBr and H_2O. These high temperature lamps use hard glasses with 0.01 to 0.05 wt.% Na_2O. Oxygen gettering is also important; use of phosphorous was reported by Beschle[6] for this system.

Incandescent-replacement lamps must be safe. The operating pressures are high. Because hard glass has poorer thermal shock resistance than fused quartz, substantially more capsules fail by bursting at end of life. The lamps' designers assumed that capsule failures could not be prevented totally, even with fused quartz, and must be contained by the outer envelope or other closure. The basis of their design[7] was to use thinner than normal capsule walls, less than 0.9 mm, and thick-walled outer bulbs, at least three times thicker than the capsule walls. Tests showed that the impacts by the low-mass glass shards had much lower energy than before. Although the initial lamps were Teflon-coated for added safety, current bulbs are uncoated. Improvement in bulb wall uniformity had led to uncoated lamps which contained as well as the original Teflon-coated products. The continued safety of these lamps is monitored in life tests; none have failed to contain at the end of life nor has an explosion in the field been reported. The importance of the safety issue requires continued and improved evaluation as more energy is concentrated into high temperature lamp envelopes.

III. Fluorescent Lamps

Phosphors

Current trends in fluorescent lighting are described in a review by Jack and Vrehen[8]. Among the materials advances summarized is the use of blends of three rare-earth-activated phosphors for high efficacy, high color-rendering lamps. These blends are more resistant to ultraviolet radiation than are standard halophosphates and lumen maintenance in lamps is much improved. Highly loaded lamp designs such as compact-fluorescent replacements for incandescent lamps and thin, 25 mm diameter, linear lamps become practical. Adjustment of the triphosphor blend ratio permits a variety of color temperatures and color rendering indices. The main disadvantage of the triphosphors is high cost.

Alternatives to the high-cost triphosphors are widely sought. One is to use two phosphor layers; a first layer of halophosphate phosphor and a second thin layer of triphosphor blend with as little expensive phosphor as possible. An optimum phosphor coating absorbs most of the UV without reflecting any back to the discharge. This optimum thickness will depend on phosphor optical properties and particle size. Sbules and Klatt[9] have applied Monte-Carlo techniques to calculate UV absorption and reflection for a variety of coating thicknesses, particle sizes and distributions. Hope[10] uses a Kubelka-Munk two-channel model to predict the spectral distribution and lumen output of blended phosphor lamps. His predictions are for a homogeneous layer, but phosphor particles can segregate during coating to seriously alter the local optical properties[11]. A dual phosphor layer lamp with reduced powder weight was described in a patent by Paynter, et al.[12]. Small particle size, 4 to 8 μ, halophosphates are a key to this invention.

Inexpensive phosphors such as $ZnSiO_4$:Mn would be alternatives to the rare-earth-based phosphors except for their poor maintenance. Phosphors deteriorate due to color centers formed by UV radiation, by ion bombardment and by reactions with mercury, impurity gases and glass. Lumen maintenance in lamps can be improved by various coatings. Fine alumina and silica coatings on the bulb or alumina blended with the phosphor are commonly used. A bulb coating of $BaSO_4/SiO_2$ is reported[13] to prevent UV-solarization of the glass, prevent mercury amalgamation with glass components and to reflect UV back to the phosphor. Seuter and van den Boom[14] disclosed using very thin continuous layers of transparent oxides such as yttria to prevent glass discoloration. In combination with triphosphors, the yttria coating on glass improved lamp maintenance with reduced phosphor powder weights. Others coat the phosphor. Sigai and Flaherty[15,16] describe thin layers of yttria and alumina on calcium halophosphate and zinc orthosilicate phosphor coatings which improve lumen maintenance. It is probable that the work in protective coatings will reduce cost while maintaining or improving lamp quality.

Fluorescent Lamp Electrodes

Life of fluorescent lamps is determined by barium loss from the oxide-coated electrodes. Bhattacharya and Hlahol[17] presented the results of LIF studies of Ba loss. They found an increased loss during the anode half-cycle as the hot spot temperature rose due to ion bombardment. Loss was found to depend on externally applied electrode heat, lamp current, and ballast characteristics. From a thermal model, Soules, et al.[18] calculated electrode temperature profiles for a variety of conditions and compared them to temperatures measured by micro-pyrometry. The calculated hot spot locations and profiles agreed well with experiment. Hot spots occur on the uncoated coil at the edge of the emission mix for externally heated electrodes, but the hot spot moves onto the emission mix with the heater disconnected. The effects of changing the wire resistance and emitter were also predicted. These studies should aid in the design of longer-lived lamps.

IV. Metal Halide HID Lamps

Materials and chemical problems in metal halide lamps are manifold. Fused silica arc tubes and associated impurities react with the metal halides to degrade performance. The electrodes emit poorly, react chemically and often deform or erode catastrophically. Molybdenum foil seals are the bane of manufacturing and a common source of arc tube failures. Chemical segregation in the discharge causes color separation. The sum of these problems is often relatively poor lumen maintenance, color variations both initially and over life, flicker and sometimes early and dramatic failure. Despite these problems, applications of metal halide continue to grow because of its relatively good color.

Recent reports on metal halide lamps are abundant and together with papers at this meeting will add to our understanding of the chemical problems. This paper will address two topics: fused silica and other arc tube materials and electrode emitter replacement.

Arc Tube Materials for Metal Halide Lamps

Fused silica has many of the ideal properties for metal halide arc tubes: a high band gap for transparency, good low-temperature strength, excellent thermal shock resistance and low electrical conductivity. It is less than ideal in other characteristics: relatively low melting temperature, low thermal expansion compared to the metal inleads, reactivity with ScI_3 and permeability to small ions and gases. A discussion of other possible arc tube materials is appropriate before returning to these and other problems of fused silica.

Alumina arc tubes have received extensive testing with several metal halide chemistries including $AlCl_3$[19], $NaI-ScI_3$[20] and $NaX-SnX_2$[21]. Although many properties are excellent, including vapor pressure and melting point, it reacts somewhat with ScI_3. More difficult are the reactions with alumina's normal complementary components, niobium inleads and ceramic seal materials. In our experiments, we have often experienced erosion of the niobium inleads with transport to the electrode tip. In early $NaI-ScI_3$ lamps, made with calcium aluminate frit seals, ScI_3 was immediately replaced by CaI_2 and BaI_2 in the vapor. Less reactive inleads of Mo/Al_2O_3 or W/Al_2O_3 cermets[22] and Ln_2O_3-based seals[23] have been developed. With these components, extended lives with reasonable performance are known for $NaI-ScI_3$[23] and particularly for the $NaX-SnX_2$[20] system. At this point, no alumina-based system is used in commercial lamps. However, the steady rate of patent publication suggests eventual success, probably with less reactive halides such as Na, Tl and In.

Perhaps success lies in other materials. Those with the most attractive material properties for $NaI-ScI_3$ lamps are BeO and Sc_2O_3. Some work has been done to develop transparent BeO[24], but its application will be severely limited by the cost and safety hazard. Little is reported on Sc_2O_3 as an arc tube, probably because it is extremely expensive. We tested sintered Sc_2O_3 by adding small pieces to standard $NaI-ScI_3$ lamps and found it to be compatible. Yttria ceramic was also tested in this manner and found to quickly remove scandium from the discharge. For this reason, it is doubtful that the recently developed $Y_3Al_5O_{12}$ ceramic[25] will be useful with the reactive Sc and Ln halides. Non-oxide ceramics or protective coating may be successful in these cases.

Fused silica has been improved for application in metal halide lamps. Metallic and water impurity levels have been decreased, driven in part by the needs of the semiconductor industry. One impurity which affects visible and U.V. transmission is iron. Concentrations above about 1 ppm cause ferromagnetic iron precipitates and darkening in the annealed product[26]. High OH-content is also a problem. Van der Steen[27] contributed greatly to understanding the system SiO_2-H_2-O_2. Most important was his identification of the reaction with hydrogen in the melt:

$$2SiO_2 + H_2 \rightleftarrows \equiv SiOH + \equiv SiH.$$

Both products can be identified by FTIR spectroscopy. Wet and reduced silica can be represented by addition of H_2O or SiO to the equilibrium. The result is a change in the concentration ratio [OH]/[H] from 1.0 for stoichiometric silica to higher values for wet silica and lower values for reduced silica. The concentrations of [OH] and [H] can be calculated from the FTIR absorbances of H_2-baked or H_2-melted silica and measured extinction coefficients. The degree of wetness or reduction was estimated from our measurements on experimental and commercial fused quartz tubes. The [OH]/[H] ratios ranged 1.3 to 0.25, or from 16 ppm [H_2O] to 50 ppm [SiO], respectively. Fused quartz used for metal halide lamps appeared reduced, 15 ppm [SiO]. The bad effect of wet quartz on metal halides is well known; that of oxygen deficiency is unknown.

One common problem in metal halide lamps is electrolytic sodium loss through the fused silica. There is some disagreement about the exact mechanism as proposed by Waymouth[28-30], but it is probable that conduction through the silica is rate determining. Sodium conductivity depends on the electric field, Na^+ concentration and diffusivity. Modification of the chemistry can affect the diffusivity of Na^+ in fused quartz. Aluminum content and the aluminum-to-alkali ratio are known to affect conductivity[31], and the effect of OH-content in sodium loss was noted by Nagel[29]. Perhaps some chemical modification of fused quartz can minimize this problem and extend life.

Non-Thoria Electrodes

Thorium oxide remains as the most commonly used emitter in metal halide lamps. Its removal is desirable because it presents a slight radiological health hazard and is reactive with ScI_3. Various metals, metal oxides and other compounds have been considered as a replacement.

Scandia has been proposed often[32,33] and used in Japan for some NaI-ScI_3 lamps. Although lamp performance was poor in our own early experiments with coated electrodes, this result is contrary to the Sc_2O_3-additive experiments mentioned above. These conflicting results suggest scandia's form and chemistry are critical in its use. The mode of operation of Sc_2O_3-W in metal halide lamps is unknown, but if it is similar to ThO_2-W, i.e., reduction and metal transport to the electrode tip, then its success depends on the stability of Sc metal at the tip temperature.

Other emitters of interest for metal halides include ceriated tungsten[34], yttrium and lanthanide oxides and hexaborides. Some will not be useful in lamps with the more reactive halides, except for combinations such as Ln_2O_3 in LnI_3 lamps. Development of non-thoria electrodes for lamps is underway and probability of success seems good.

V. High-Pressure Sodium Lamps

The state-of-the-art in HPS lamps is detailed in the excellent new text by de Groot and van Vliet[35]. A review of sodium reactions with alumina tubes and seals is covered there and in a review by Akutsu[36]. Studies of HPS materials problems were presented at the 1988 Symposium on High Temperature Lamp Chemistry. This review will consider three areas: arc tube fracture, high color rendering lamps and unsaturated vapor lamps.

Alumina Strength

De With[37] measured strength of polycrystalline alumina (PCA) tubes by the diametral compression fracture method. The measured strengths depended on ring diameter and length. We also measured PCA strengths by this method and thermal shock resistance by a pulsed CO_2-laser method[38]. We found that strength depended on size and distribution of the grains. Mean strengths correlated with (mean grain size)$^{-1/2}$ as expected, but

MgO-doped PCA tubes had a different correlation than MgO/Y_2O_3-doped PCA. Both materials, however, had the same correlation to maximum grain size. Dependence on (maximum grain size)$^{-1/2}$ is found when the grains represent the largest flaws[39]. The difference between the two aluminas was the bimodal grain size distribution of MgO/Y_2O_3-doped PCA, which has a larger maximum grain size for the same mean. The thermal shock resistance dependence on grain size was similar. It was concluded that attention to maximum grain size is important in arc tube designs subject to fracture.

HCRI Lamps

High color rendering (HCRI) lamps include sodium lamps with CRI above 60 and sodium lamps with other metal additives to improve color. Although 2500 K, 85 CRI lamps are much whiter, the efficacy is low and red rendition is somewhat exaggerated. Additional metal additives could improve performance. In reported examples[40], the color temperature and CRI are still relatively low. The paper by Snellgrove[41] focuses on the problems of a white Na-Cd-Tl-Hg lamp which operates with seal temperatures above 950 °C.

The problems of sodium reactions with seals and alumina in HCRI lamps have been discussed extensively[36]. In tests of seal materials in our laboratory[42], a Sc_2O_3-containing aluminate frit similar to one described by Oomen and Rouwendal[43] was the most resistant in lamps operating at over 25 kPa sodium pressure.

USV Lamps

Unsaturated vapor (USV) HPS lamps were first disclosed by Schmidt[44]. Study of the technology has been renewed[45-48] because voltage rise due to sodium loss is prevented at the end of life. USV presents a technical and manufacturing challenge since sodium loss must be limited to tens of micrograms over lamp life. Low oxygen partial pressure is maintained through careful processing and choice of materials. Emissive materials and seals are particularly important. $BaCaWO_6$ emitters, used in many HPS lamps, lead to rapid sodium loss and oxides of Th, Y, or Ln are generally employed instead. Investigation of these lamps is active in most lamp companies, but few have announced or sell products.

Summary

New lamp development continues vigorously on all major lamp types. In most, the limits of materials are being tested. Lighting scientists will likely solve the material problems in today's lamps. However, revolutionary lamp developments may await breakthroughs from outside the lighting community. Our job will be to recognize the opportunities.

Acknowledgements

I would like to thank my colleagues at GTE for their assistance in gathering material during preparation of this paper. Particularly helpful were Ernest Dale, William Keeffe, Stephen Kimball and Dr. Richard Snellgrove in their constructive criticism of the manuscript.

References

1. Rosner, D. E. Proc. Symp. High Temp. Lamp Chem. II, E. G. Zubler, Ed., The Electrochemical Society, p. 111, (1988).
2. Würster, E. and Heidemann, A. U.S. Patent 4 508 991, 1985.
3. Kiesel, R. and Blum, K. U.S. Patent 4 777 404, 1988.
4. Dittmer, G. and Niemann, U. Philips J. Res. 42, 13 (1987).
5. Kimball, S. F., Audesse, E. G. and Griffin, R. M. U.S. Patent 4 550 270, 1985.
6. Beschle, M. 4th Int. Symp. Sci. and Tech. Light Sources; Karlsruhe (1986).
7. Gagnon, P. R. U.S. Patent 4 598 225, 1986.
8. Jack, A. G. and Vrehen, Q. H. F. Philips Tech. Rev. 42, 342 (1986).
9. Soules, T. F. and Klatt, W. A. J. Illum. Eng. Soc. 17, 92 (1988).
10. Hope, L. L. 5th Int. Symp. Sci. and Tech. Light Sources, York (1989).
11. Dale, E. A. unpublished results.
12. Paynter J., Gillooly, G. R., Soules, T. F. and Hoffman, M. V. U.S. Patent 4 806 824, 1989.
13. Jönsson, G. Eur. Patent Application Pub. 0289474A2, Applic. 88850132.7, 1988.
14. Seuter, A. M. J. H. and van den Boom, P. F. J. U. S. Patent 4 544 997, 1985.
15. Flaherty, J. M. U.S. Patent 4 459 507, 1984.
16. Sigai, A. G. and Flaherty, J. M. U.S. Patent 4 670 688, 1987.
17. Bhattacharya, A. K. and Hlahol, P. G. 4th Int. Symp. Sci. and Tech. Light Sources; Karlsruhe, (1986).
18. Soules, T. F., Ingold, J. H., Battacharya, A. K. and Springer, R. H. J. Illum. Eng. Soc., in press.
19. Speros, D. M. and Smyser, W. E. U.S. Patent 3 586 898, 1971.

20. Akutsu, H., Yamazaki, H., Iwama, K. and Saito, N. U.S. Patent 3 911 308, 1975.
21. Brown, K. E., Chalmers, A. G. and Wharmby, D. O. J. Illum. Eng. Soc. 11 106 (1982).
22. Hing. P. Science of Ceramics 9 135 (1977).
23. Yamasaki, H., Kitagawa, M. and Tahastu, K. Jap. Patent 61-245458, 1986.
24. Anderson, N. and Richter, J. Final Report R-ILC-73-18 on Contract N00014-71-0384 by ILC Technology, (1973).
25. de With, G. Philips J. Res. 42 119 (1987).
26. Griscom, D. L., Friebele, E. J. and Shinn, D. B. J. Appl. Phys. 50 2402 (1979).
27. Van der Steen, G.H.A.M. and Papanikolau, E. Philips Res. Rep. 30 309 (1975).
28. Waymouth, J. F. et al. Illum. Engr. 62, 214 (1967).
29. Nagel, F. Proc. Symp. High Temp. Lamp Chem. II, E. G. Zubler, Ed., The Electrochemical Society, p. 18 (1988).
30. Keeffe, W. M. and Meyer, V. D. to be presented at 5th Int. Symp. Sci. & Tech. of Light Sources, York (1989).
31. Varshneya, A. K., Bhargava, A. and Jain, V. Abstracts of Papers, 89th Annual Meeting of the American Ceramic Society, Pittsburg, PA, American Ceramic Society, Westerville, OH, 1987; Abstract 66-G-87.
32. Rigden, S. A. R. U.S. Patent 4 232 243, 1980.
33. Davenport, J. M., et al. U.S. Patent 4 574 219, 1986.
34. Wang, J. Z. U.S. Patent 4 678 718, 1987.
35. de Groot, J. and van Vliet, J. The High-Pressure Sodium Lamp; Kluwer Technical Books B.V.: Deventer-Antweip, (1986).
36. Akutsu, H. Lighting Research & Technology 16 73 (1984).
37. de With, G. J. Mat. Sci. Letters 3 1000 (1984).
38. Shinn, D. B. and Tiernan, R. J. unpublished results.
39. Alford, N. McN., Birchall, J. D. and Kendall, K. Nature 330 51 (1987).
40. Saito, S., Tominage, J. and Tsutsumi, N. Jap. Patent 63-2243 (1988).
41. Snellgrove, R. A. to be presented at 5th Int. Symp. Sci. & Tech. of Light Sources, York (1989).
42. Schoene, C. M. unpublished results.
43. Oomen, J. J. C. and Rouwendal, J. W. Proc. Symp. High Temp. Lamp Chem., E. G. Zubler, Ed., p. 291 (1985).
44. Schmidt, K. U.S. Patent 3 248 590, 1966.
45. Furukubo, H., Yasaburo, T. and Takatsuka, K. U.S. Patent 4 156 550, 1979.
46. Okada, T., Yokome, H., Kosasa, A., Habu, T. and Yamamoto, Y. U.S. Patent 4 755 721, 1988.
47. Yokome, H., Okadu, M., Kosasa, O. and Yamamoto, Y. Jap. Patent 62-82640, 1987.
48. White, P. J. and Wyner, E. F. J. Illum. Eng. Soc. 14 95 (1985).

REVIEW OF ELECTRODELESS DISCHARGES FOR LIGHTING

D O Wharmby

Lamp Research, THORN Lighting Limited, Light Sources Division,

Melton Road, Leicester LE4 7PD, U.K.

1. INTRODUCTION

Electrodeless discharges (ELD) - discharges with no internal electrodes - offer a tantalising prospect for the light source manufacturer looking to overcome the disadvantages of existing light sources. Shortly after the first observations of ELD by Hittorf and by Thompson, Tesla (1891) had the idea of creating a high frequency field in a room; then "the mere suspension of the tubes in the room would afford the desired illumination". Almost a century from that date, we still have no commercial general lighting sources based on the electrodeless principle. There are of course a variety of ELD available for special purposes such as ink curing (Philips 1983) and spectroscopy (Boumans 1987). What are the reasons that have so far prevented the commercial application ELD lamps for lighting?

There has been enormous activity in the last three decades, paralleling the developments in semiconductors which are an essential ingredient. In the past 15 years alone there have been over 200 patent applications on ELD of various types. Many come from Japan, with Matsushita, Mitsubishi and Toshiba particularly active. Over 100 applications have been directly concerned with sources for lighting.

Electrodeless discharges appear to offer the following technological possibilities as sources for general lighting:

* absence of electrodes, their losses, and especially their failure mechanisms

discharges are possible (figure 1 gives some examples) and appear in the patent literature: the primary can be inside, outside, or otherwiase disposed relative to the plasma secondary, with and without a ferromagnetic core. All that is required is that enough of the changing primary flux cuts the secondary to generate the maintenance voltage V_m for a stable discharge. At an angular frequency ω, the changing flux required to generate this field is proportional to ωI_p, and so to ωI_s, where I_p and I_s are the primary and secondary currents. To maintain the discharge at low frequencies, a large value of I_p, and thus of power $I_s V_m$ is required: in contrast to E discharges, low frequency H discharges can only be maintained at high power. In Babat's remarkable research H discharges of up 100 kW were investigated.

There has been a considerable debate in the literature as to whether H discharges ever exist independently of E discharges. For any coil configuration, the voltage across the ends of the coil is always greater than the azimuthal voltage generated by the changing flux (Contaxes and Hatch 1969). Townsend and Donaldson (1928) showed that unless special precautions are taken, H discharges are always started by the voltage across the ends of the coil: a useful benefit for lighting applications. In steady state inductively coupled discharges, the plasma is driven by a combination of E and H fields, relative contributions depending on a number of factors such as frequency and impedance characteristics of the generator and discharge etc. I have observed that inductively coupled, high pressure discharges (at 67 MHz) can switch discontinuously between a number of modes such as diffuse, ball and ring: these can be unstable particularly if metal halides are involved. A study of high pressure H discharges of interest in lighting is given by Anderson et al. (1985), and a useful review of the related inductively coupled plasmas is given by Eckert (1974).

* release from some of the well-known constraints on lamp design, allowing innovative light sources to be created

* use of metal-halide and other reactive doses such as permanent gases and oxyhalides

* low pressure discharges which produce white light directly in the visible region

* compact lamps with integral electronics

* extremely high efficacy: 171 lm/RF watt has been measured in a high-power high-pressure metal halide ELD (Hollister 1986)

* very long lamp life, particularly for frequently switched lamps

* a lamp with a cheap replaceable discharge tube

2. CLASSIFICATION OF ELECTRODELESS DISCHARGES

Babat (1942) made a classification of ELD, extended here, which has proved extremely useful in understanding the important features. The theory of high frequency discharges is reviewed by Francis (1960).

E DISCHARGES

The simplest form of E discharge is a gas filled tube between the plates of a capacitor. Displacement current must flow through the capacitance of the (glass) wall, the dc sheaths and the fluctuating ac sheaths at the edge of the plasma. The inevitably low values of this capacitance means that it is difficult to dissipate much power in an E discharge except at high frequencies. Nevertheless E discharges for lighting appear in the patent literature, for example Proud et al. (1979). E discharges of this type are closely analogous to the RF discharges with internal electrodes reviewed by Godyak (1986).

H DISCHARGES

In an H discharge the plasma forms the single turn secondary of a transformer, the primary being driven from the power source. Such discharges are variously referred to as 'inductively coupled', 'solenoidal electric field' and 'divergence free'. Many variants of H

MICROWAVE DISCHARGES

At microwave frequencies, where the wavelength becomes comparable with the dimensions of the discharge tube, the discharge is excited by both the E and B components of the field. A light source at 915 MHz has been extensively covered in the patent literature by the GTE group (Haugsjaa et al. 1975). The discharge is arranged to terminate a transmission line. Arrangements must be made so that a high enough field is available for starting, that the generator can survive the reflected power before starting, and that after starting the plasma presents the characteristic impedance to the line. Judging by the content of the patent literature, this can be quite tricky. Sodium scandium iodide discharges of 90 lm/RF watt at 50W in such devices were reported at a previous symposium (Haugsjaa 1979). The commercialisation of ELD for lighting operating at 915 MHz and above would seem unlikely at the present time because of the cost and limitations of the semiconductor circuitry.

TRAVELLING WAVE DISCHARGES

Plasmas can be created by travelling waves in waveguides and slow wave structures (Marec et al. 1983). A rather useful form appears to be the surface wave discharge, well known in the scientific literature, in which the wave energy forms the plasma and the plasma itself is the structure along which the wave is propagated. Figure 2 shows a typical surface wave plasma device (Marec et al. 1983, Levy and Berman 1985). The RF power is fed to coupling electrode. The electric field ionizes the gas in the gap and a surface wave propagates out of the gap along the boundary of the glass and the plasma. Surface waves are electromagnetic wave modes propagating at frequencies well below the electron plasma frequency and at velocities only a fraction of the velocity of light.

Advantages of surface waves seem to be their predictability, the variety of devices which can be used to launch them (Moisan et al. 1982), the wide range of frequencies usable (MHz to GHz), the high field strength in the gap permitting easy starting, the evanescent nature of the radial electric field and the fact that in some conditions the power flow can be near the surface of the plasma. This can lead to enhanced excitation of atoms near to the wall (Moisan and Zakrzewksi, 1986). Radiation transport in mercury rare-gas surface wave discharges may therefore be different from that in fluorescent lamp discharges, although it not obvious that this will lead to more efficient generation of UV. Levy and Berman (1986) claim up to a 37% improvement in using surface wave excitation at 530 MHz compared with mains excitation of the same 25 mm diameter 15W fluorescent lamp. This is hard to understand, since the efficiency of UV generation of the fluorescent discharge is already in the region of 70%, but it certainly would bear further investigation.

3. SOME PRACTICAL DEVICES FOR LIGHTING

Although there are no commercial ELD yet available, some of the variants based on H discharges are clearly quite practical (figure 1). These have all evolved in one way or another from Thomson (1927) and Bethenod and Claude (1933), who described air-cored and ferromagnetic cored H discharges respectively. The principle proponents are Anderson of GE, Hollister and a group from Philips, Holland; all have worked with compact lamps using mercury/rare gas discharges, mainly in a bulb looking like a household bulb in shape. Other variants on these themes appear in Soviet (Volochov 1983), Chinese (Hsu and Chen 1983) and Japanese research: Matshusita (1984) appear to be making patent applications for a variety of versions in novel formats (figure 1d).

Anderson (1975, for example) has largely followed the notion of using a toroidal ferrite core either inside the bulb, or outside it in a reentrant cavity (figure 1a). The core is driven by a primary winding from a simple oscillator. The advantage of this geometry is that there is little flux leakage. At the rather low frequencies chosen, a large peak value of B (magnetic field) is needed to produce the required discharge maintenance field; with the best compromise conditions the ferrite is driven rather hard with resulting high losses. This problem, and the high cost of the ferrite components appear to be reason why this development seems to have ceased. Nevertheless Anderson has made a notable contribution to the development of ELD.

Wishing to avoid the disadvantages of the ferrite core, Hollister (1975) uses a copper solenoid of about 10 turns in a reentrant cavity in the bulb at a frequency of typically 4 MHz (figure 1b). Substantial commercial development has been put into this concept on at least two occasions by companies named Litek and Diabolo: the Diabolo lamp operates at the ISM frequency of 13.56 MHz. A photograph of a Litek lamp (anonymous 1976) clearly showed one of the difficulties, in that the circuit, far from being hidden in the cap of the lamp, occupied a sizeable box beneath it. The overall system efficacies are also rather marginal at 40-50 lm/W.

More recently, the Philips group has worked with a device superficially similar in geometry to the Hollister concept, but having a cylindrical ferrite core and operating at 2.65 MHz. This appears to be the most realistic concept yet for a number of reasons. Firstly, Postma and van Vliet (1985) quote 92 lm/RF watt. Secondly there seems to a serious effort to solve the radio frequency interference problem, with values of radiated field quoted that are quite close to those required by the regulations. Thirdly there is evidence in their patent applications of attention to the sort of details needed in a commercial product.

4. PROBLEMS RESTRICTING EXPLOITATION OF ELD

In the early developments phosphor maintenance was a problem, a loading of about 1000 W/sq.m being required for a typical compact incandescent lamp replacement. Loadings of this order or more are now commonplace in fluorescent lamps, thanks to rare-earth phosphors.

Safety is an issue that must be considered, as with any equipment operated at radio frequencies. Imagine a 10W lamp with a diameter of 10cm. If all the 10 W input were radiated as RF power isotropically, the power density at the surface of the bulb is only about 30mW/sq.cm. It should be easy in any practical device to keep below the recommended levels of around 1-10 mW/sq.cm. Statutory safety regulations will not therefore prevent the development of ELD.

Radio frequency interference (RFI) is another matter. International agreements set the maximum levels for electromagnetic disturbance limits from unintentional radiators to roughly 30-40 dB (μV/m) at 10m, at the frequencies that have been used. This level is set in IEC standards to avoid interference with communications equipment. These levels will be difficult to meet with any practical shielding.

The obvious option is to use the internationally agreed ISM frequencies, which are set aside for non-communications use of the radio spectrum: unlimited radiation is permitted in these bands. There are relatively few bands available and some are still being discussed. For example, because of conflict with mobile communications, there is no clear agreement in sight in Europe for a band equivalent to the 915 MHz band in the USA. In order to keep the door open for potential developments in ELD it is important for the lighting industry to present a unified lobby at International level. A useful harmonic sequence of ISM bands are those at 13.56, 27.12, and 40.68 MHz. The width of these bands is 30, 300 and 40 kHz, respectively. It is unlikely that free running oscillators will be stable enough to operate within the bands, considering the change in plasma impedance between the on and off states of an H discharge.

I notice that few of the devices referred to in section 3 are operated in the ISM bands. Do the inventors have some means for avoiding RFI? Many patents mention thin film semiconductor coatings (indium or tin oxide) almost as a matter of course. Obvious though this is as a solution, such films, if they are to be sufficiently transparent can only offer 20-30 dB screening, and a good deal more is needed for operation other than at the ISM frequencies. Other approaches to screening involve special forms of inductive loop in which the dipole contribution cancels at large distances leaving only the quadrupole component in the far field (eg Buhrer 1979).

Electronics still presents the major challenge. As in all lamp electronics the requirements of efficiency, compactness, reliability and low cost are difficult to achieve simultaneously. A typical RF circuit may comprise an oscillator and amplifier (or perhaps a free running oscillator) operating from a DC power supply. Some idea of the magnitude of the challenge can be seen by assuming that RF and DC stages can both be made 80% efficient. Then the 92 lm/RF watt lamp mentioned above, which operates at 13W (RF), would take 20.3 W from the mains, giving an overall efficacy from the mains of 59 lm/W, and leaving 7.3W to dissipate in the compact power supply in the base. Class A, B and C circuits for RF power generation are problematical on account of their theoretical limiting efficiencies. Class D and E circuits both have theoretical efficiencies of 100% (Lutteke and Raets 1984), with the actual limits being set by transistor switching characteristics. Little seems to have been done by semiconductor manufacturers to optimise devices specifically for driving lamps, despite the potential for large volume business.

5. CONCLUSIONS

A huge amount of work has been done on ELD in recent years, with the most practical outcome being compact sources based on H discharges of the fluorescent type at frequencies in the 0.2 to 20 MHz range. There are signs that discharge efficacy can approach 100 lm/RF watt. The major technical problem is to achieve high efficiency electronics leading to high system efficacy and reliability. Problems of RFI can be overcome if

ISM frequencies are used. The lighting industry is urged to present a united front on further standardisation in this area.

I am grateful for colleagues in Thorn Lighting and Thorn EMI Central Research Laboratories for useful discussions.

6. REFERENCES (dates attached to patents are first filing dates)

Anderson J M, 1975, US Patent 3,987,335, 18 Dec.
Anderson J M, et al., 1985, Lawrence Berkeley Lab. Report, LBL-19035
Anonymous, 1976, Ltg. Des. Appl., April, 48-9
Babat G I, 1942, JIEE, 94, 27-37
Bethenod J and Claude A, 1933, US Patent 2,030,957, 29 Dec.
Buhrer C F, 1979, US Patent 4,245,179, 18 Jun.
Boumans P W J M, 1987, 'Inductively Coupled Plasma Emission Spectroscopy', (New York:Wiley)
Contaxes N and Hatch A J, 1969, J. Appl. Phys., 40(9), 3548
Eckert H U, 1974, High Temp. Sci., 6, 99-134
Francis G., 1960, 'Ionization Phenomena in Gases', (London:Butterworths)
Levy and Berman, 1986, European Patent Appl. 0 225 753, 19 Nov.
Godyak V A, 1986, 'Soviet Radio Frequency Discharge Research', (Falls Church, Va. USA:Delphic Associates)
Haugsjaa P O et al., 1975, US Patent, 3,943,401, 21 Apr.
Haugsjaa P O et al., 1979, 'High Pressure Microwave Discharges as Light Sources', 3rd Int. Symp. Incoh. Light Sources, Enschede, paper 9
Hollister D D, 1975, US Patent, 4,010,400, 13 Aug.
Hollister D D, 1986, SPIE, 692, 170-7
Hsu H C and Chen G X, 1983, Int. J. Electronics, 55 (6), 829-41
Lutteke G and Raets HC, 1984, 15th Annual IEEE Power Electronics Specialists Conf., Gaithersburg Md., USA, 54-61
Marec J et al., 1983, 'Microwave Discharges' in 'Electrical Breakdown and Discharges in Gases' ed. E E Kunhardt and L H Luessen, NATO ASI Series, 89b, 347-82
Matshushita Electric Works, 1984, Japanese Patent Appl.,J 1096-649A
Moisan M, Shivarova A and Trivelpiece A W, 1982, Plasma Phys., 24, 1331
Moisan M and Zakrzewski Z, 1986, 'Plasmas Sustained by Surface Waves at Microwave and RF Frequencies' in 'Radiative Processes in Discharge Plasmas' ed. J M Proud and L H Luessen, NATO ASI Series 149, 381-430
Philips R, 1983, 'Sources and Applications of UV Radiation', (New York: Academic Press)
Postma P ,van Vliet J A J M, 1985, UK Patent Appl.,2 161 982 A, 10 Jun.
Proud J M and Smith R K, 1979, US Patent 4,266,166, 5 May
Tesla N, 1891, Elec. Engineer, 7, 549-50
Thomson J J, 1891, Phil. Mag. 32 (197), 321-36 and 445-64
Thomson J J, 1927, Phil. Mag. 4 (25), 1128-60
Townsend J S and Donaldson R H, 1928, Phil. Mag., 5 , 178-91
Volochov A A et al., 1983, USSR patent SU 1,241,306, 14 Dec.

Figure 1: Examples of H discharges

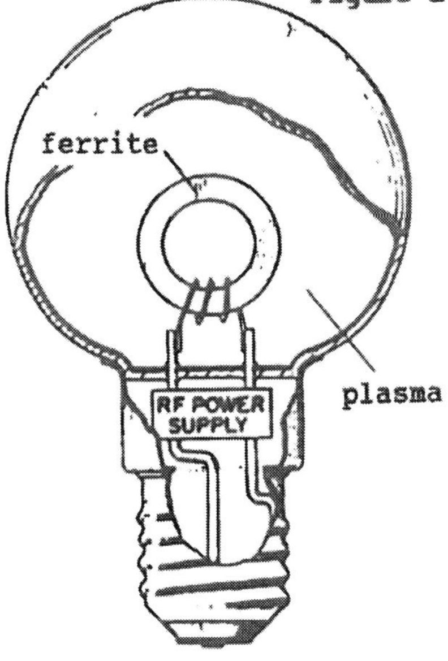

(a) Anderson (1975): toroidal ferrite core with intersecting plasma

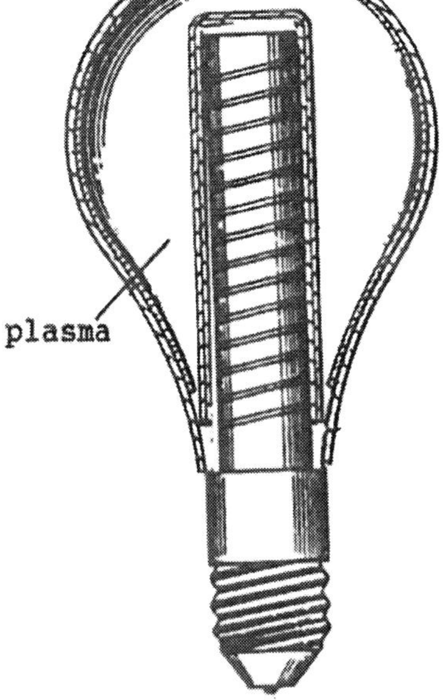

(b) Hollister (1975): air core, toroidal plasma surrounds coil

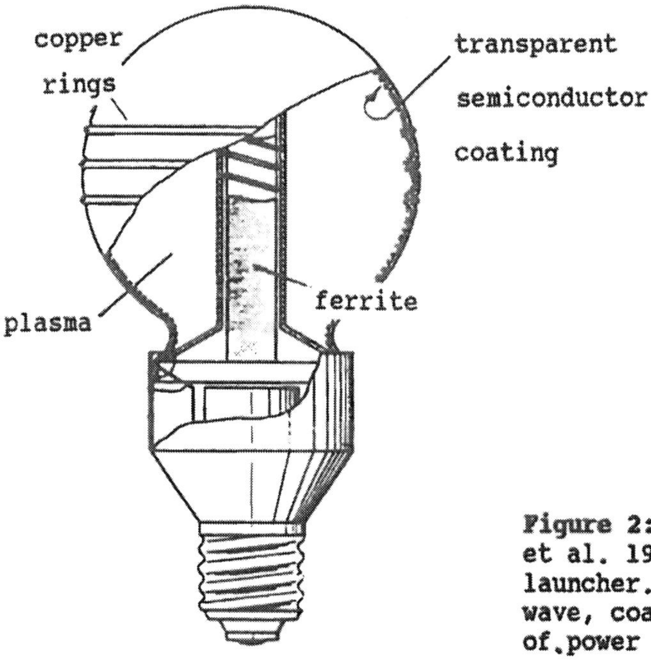

(c) Postma & van Vliet (1985): ferrite core,toroidal plasma surrounds coil. Note interference suppression rings and coatings.

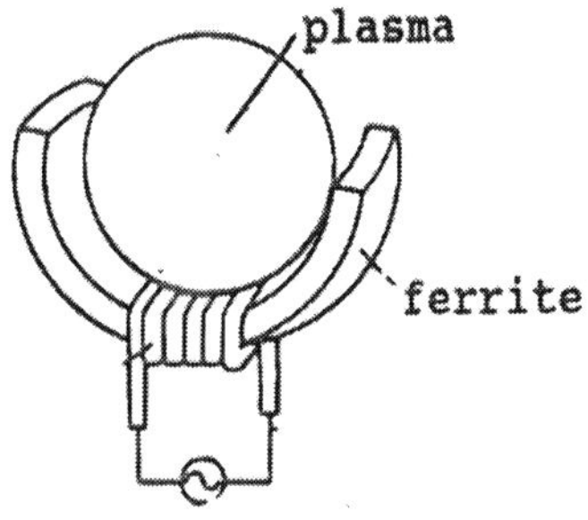

(d) Matsushita (1984): ferrite core generates shaped magnetic field

Figure 2: Surface wave discharge (Moisan et al. 1982) section through cylindrical launcher. Field in gap launches surface wave, coaxial transmission line to left of.power feed is impedance matcher.

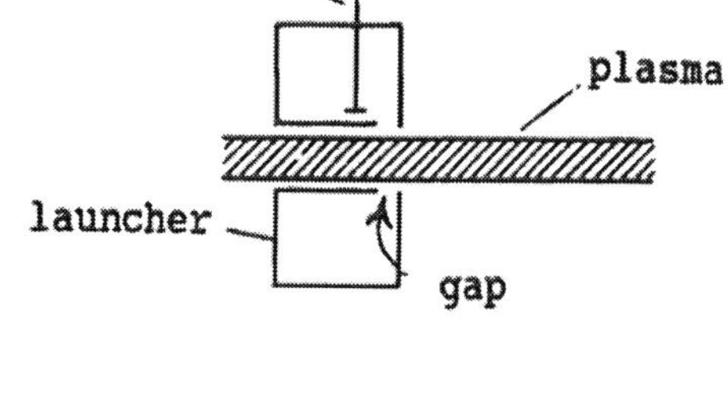

Power supply for 2 kW halogen lamp with continuous power control and sinusoidal input current

Manfred Albach

Philips GmbH, Research Laboratory Aachen, Postfach 1980, D-5100 Aachen, W.-Germany

In this paper a power supply for the continuous control of a 2 kW halogen lamp for light cooking applications is presented.

In the lower power range dimming of these lamps is normally realized by phase control. The average power supplied to the load is controlled by varying the conduction time within the half-cycle of the mains voltage. This control principle produces a high level of mains current harmonics and its application is therefore prohibited for heating elements and thermal appliances in the power range >200 W /1/. A second possibility for controlling the average power to the load is the multicycle control, where the ratio of the number of half-cycles of current conduction to the number of half-cycles of non-conduction is varied. This principle, however, causes flickering and high inrush currents at the beginning of every conduction period.

In light cooking applications a visible indication of the power setting as well as a constant "light impression" for all power levels is desirable. For these reasons a switched mode power supply with sinusoidal input current (fulfills all present and future regulations concerning mains current harmonics), a soft start (no inrush current due to cold lamps) and a continuous power setting from 0 to 2 kW has been designed. The power circuit (see Fig. 1) consists of two 1 kW down-converter circuits, which are operated in parallel with a switching frequency of 140 kHz, a rectifier bridge and a high frequency filter in order to reduce the conducted emissions below the limits of VDE 0871 class B.

From the control circuit Fig. 2 and the corresponding voltage waveshapes Fig. 3 it becomes clear, that the two MOSFET switches in the converters are operated with the same frequency and the same duty cycle. After switch-on the supply voltage V_s for the logic circuitry is generated. At $V_s = 8$ V a comparator switches and the reference voltage V_{ref} starts to increase with a preset time constant. By comparing this reference voltage with a sawtooth voltage the pulse width modulated driving signal for the MOSFETS is generated. Using a potentiometer a reference voltage variation and therefore a duty cycle variation of $0 \le \delta \le .95$ can be realized. This gives a continuous lamp power control in the range $0 \le P \le 1.9$ kW. Above this value the two MOSFETS are conducting permanently, thus avoiding switching losses and interference problems at full load. With two different time constants for an increasing and decreasing reference voltage, a slow turn on (soft start) and a fast turn off of the lamps is achieved. Due to a phase shift of 50% in the two driving signals ③ and ④ the 140 kHz component of the input current is compensated and therefore the HF-filter can be considerably reduced.

The mains current for 1 kW output power is shown in Fig. 4. The efficiency of the total circuit is 97.5% and the power density of this circuit including heat sinks and a small fan is 2 kW/litre.

Reference:

/1/ Disturbances in supply systems caused by household appliances and similar electrical equipment, IEC Publication 555, 1982

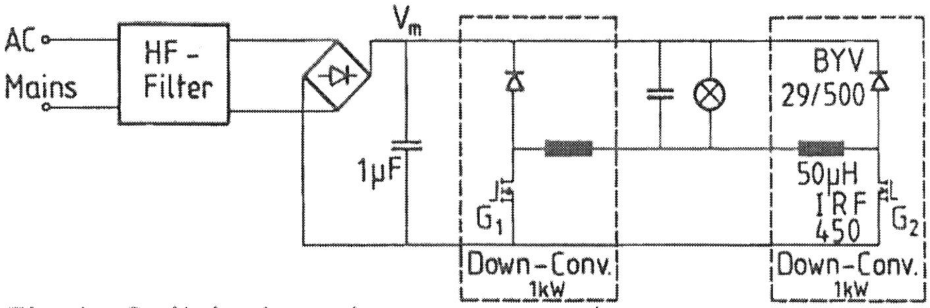

Fig. 1 Switched mode power supply

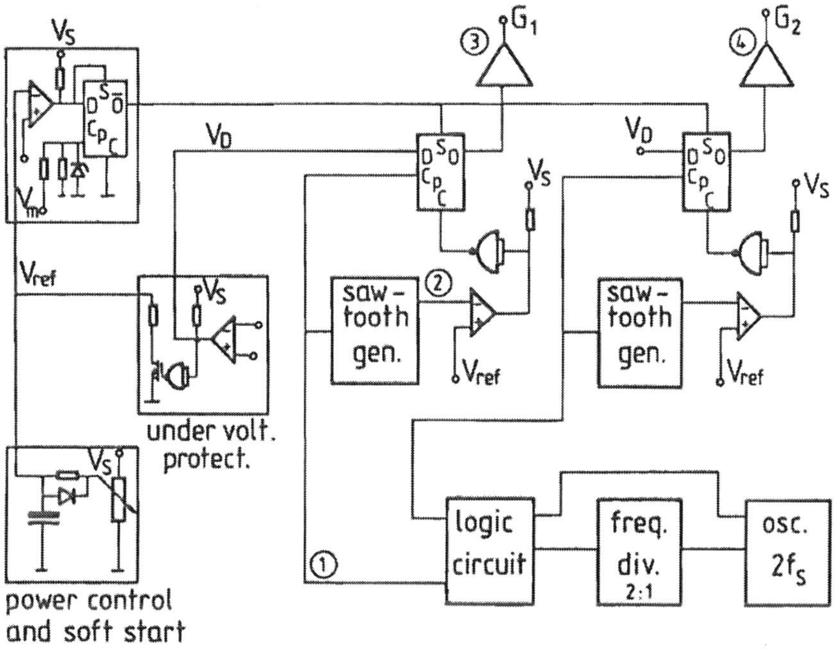

Fig. 2 Control circuit

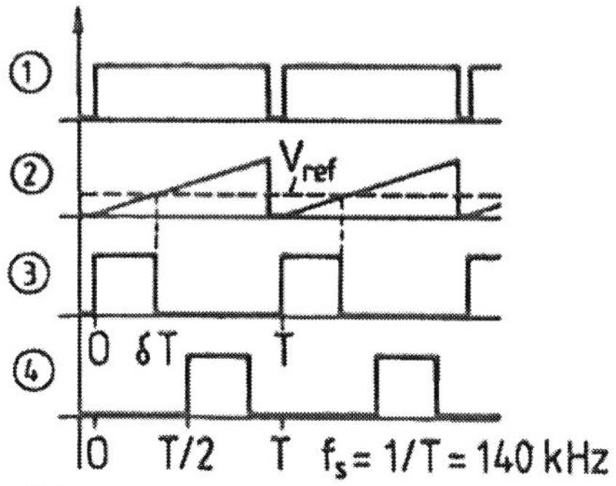

Fig. 3 Voltage waveshapes

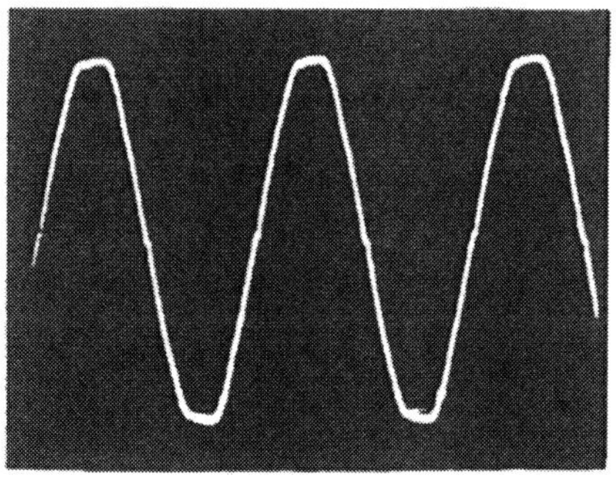

Fig. 4 Mains current (Pout = 1kW)

Impedance and Emission Properties of Capacitively Coupled Hg-Ar Low Pressure Discharges

C. Beneking
Philips GmbH Forschungslaboratorium Aachen,
Weißhausstraße, D-5100 Aachen, F.R.G.

Capacitively coupled, electrodeless low pressure discharges are of interest for future fluorescent lamp or UV source applications. With the aim to determine the physical mechanisms limiting the brightness and efficiency of these discharges, the impedance and emission properties have been studied for the example of Ar buffered discharges in the 10 to 400 MHz range. The RF power was coupled into quartz discharge vessels with a volume of $\simeq 10\ cm^3$.

For a given absorbed power (e.g. 10 W), the emission intensity and hence the efficiency increases strongly with frequency (fig. 1).

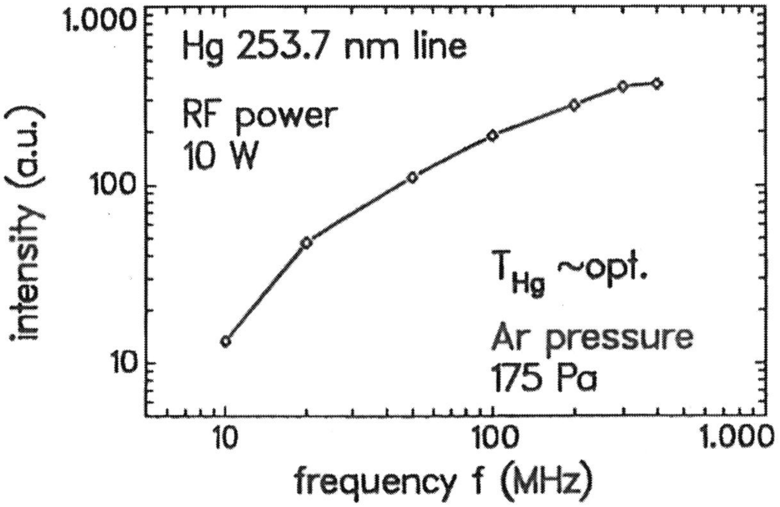

Fig. 1: Intensity of Hg 253.7 nm radiation as function of operating frequency for a 10 W capacitively coupled low pressure lamp.

This phenomenon is attributed to the varying share of power deposition between the bulk plasma ('positive column') and the non-luminous interaction regions between plasma and walls ('sheaths').

The power deposition profile is obtained from an analysis of the electrical impedance $Z = R + jX$ of the discharge. The relevant quantity is the resistive component R, because R is directly related to the dissipated power P by $P = R \times I^2$ (I discharge current).

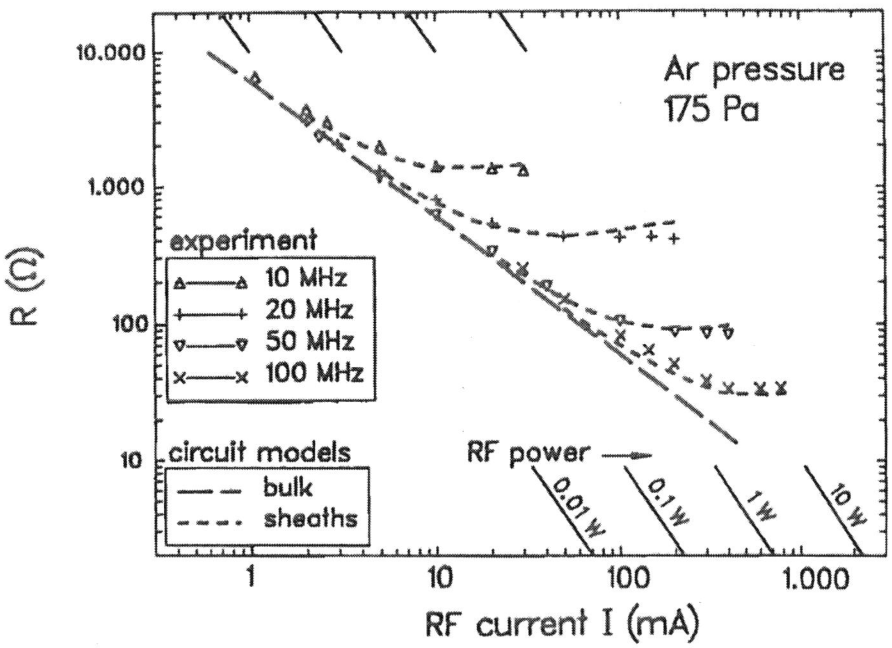

Fig. 2: Resistive impedance component R of a capacitively coupled low pressure discharge as function of the discharge current at four different frequencies. The lines of constant power are indicated for convenience.

The results shown in fig. 2 are obtained by a high resolution phase measurement between the current and voltage waveforms using a RF vector analyzer. The resistive impedance components measured at different frequencies show a common frequency-independent branch at low currents which is assigned to the bulk resistance R_b. The branches at higher currents are decreasing with frequency. They are assigned to the sheaths.

The resistive components R_b of the bulk and R_s of the sheaths are discussed with the help of equivalent circuit models. The good agreement between respective model calculations (dashed lines in fig. 2) and the experimental results gives evidence that the collisional ion motion in the space-charge fields, as expressed by the 'mobility-limited' Child-Langmuir law [1], constitutes the dominant loss process in the sheaths.

The sheath thickness is derived from the reactive component X. It decreases with frequency but is nearly independent of the current. Additional information is obtained from spatially resolved spectroscopy of the Ar excitation.

[1] R. Warren, Phys. Rev. 98 (1955) 1658

Rg_2* and RgCl* Excimer Emission from Dielectric Barrier Discharges

B. Gellert, B. Eliasson, U. Kogelschatz, Asea Brown Boveri Corporate Research, 5405 Baden / Switzerland

Xe_2* emission excited in a dielectric barrier discharge has previously been systematically investigated [1]. It was found that the efficiency reached 6-9%. Meanwhile, most of the rare gas excimers could be excited by the authors. The spectra of Ar_2*, Kr_2* and Xe_2* are shown in Fig. 1:

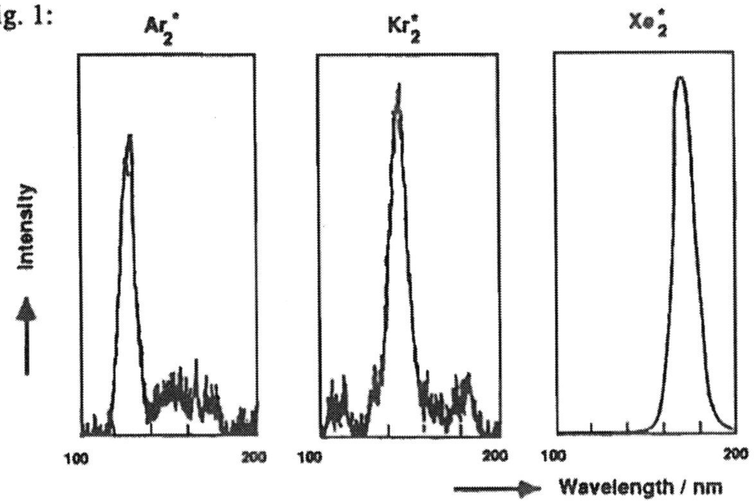

Also radiation from rare gas-halide excimers like ArCl*, KrCl*, XeCl* can be obtained easily.

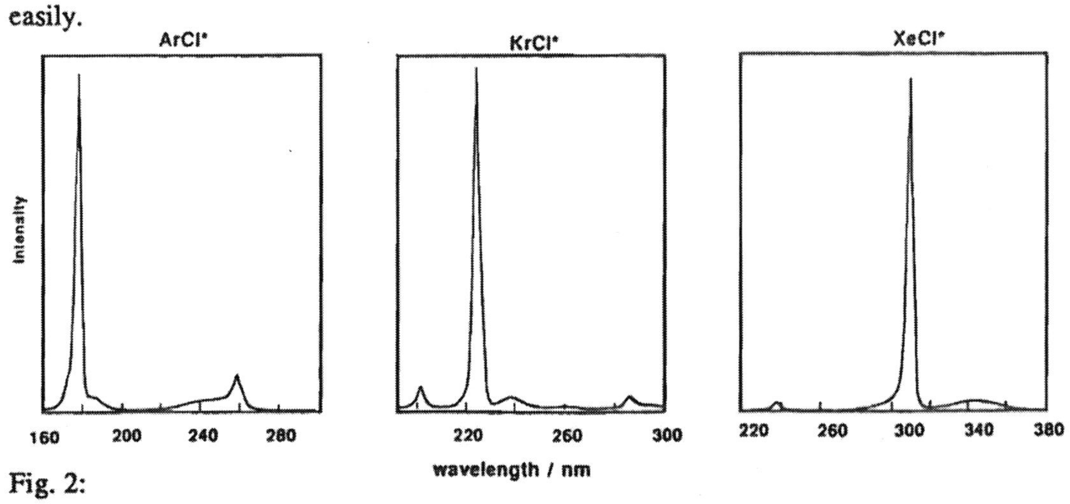

Fig. 2:

Fig. 2 shows the chloride spectra. The gas mixture contained Cl_2 in Ar and Kr or Xe. In all cases Ar was used as a buffer gas. Changing the Ar buffer gas pressure resulted in wavelength changes. Increasing the Ar content results in a strongly increased ratio of KrCl* emission compared to ArCl* radiation (Fig. 3). Even with Ar buffer gas there are discharge conditions where ArCl* emission is negligible compared to KrCl* emission. For the XeCl* radiation the results are similar. The efficiency of the new UV sources increases with increasing pressure and reaches approximately 12% at 800 mbar.

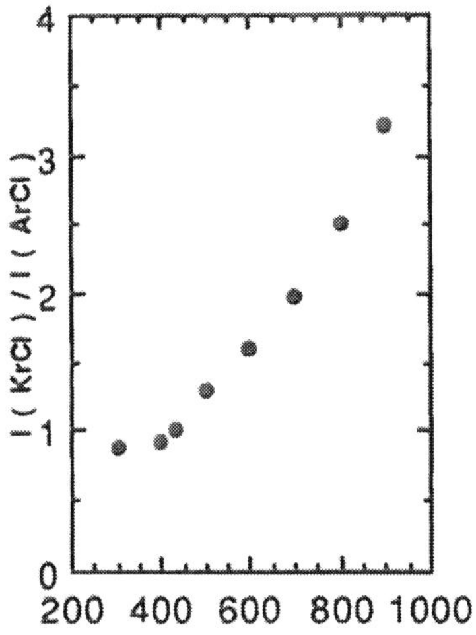

Fig. 3: p / mbar

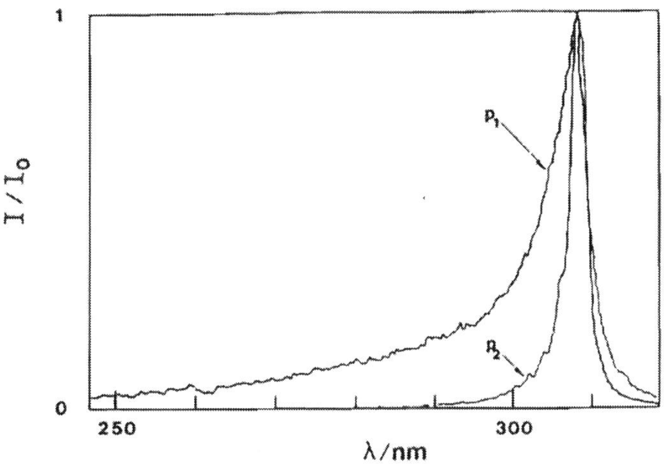

Fig. 4:

The change of the 308 nm-spectrum with pressure is shown in Fig. 4 where two spectra are compared. The broad spectrum with half width of approx. 6.6 nm can be measured at low total pressure e.g. for p_1 = 120 mbar. The higher the total pressure, the narrower the XeCl* emission at 308 nm becomes. The spectrum with half width of 2 nm corresponds to a gas pressure of p_2 = 920 mbar. Because emission at wavelengths other than 308 nm is negligible in the wavelength region up to 800 nm, similar to excimer lasers this source has great potential for applications where "monochromatic" UV radiation is required.

The described UV and VUV light sources are particularly suited for all photoassisted CVD techniques where up to now lasers, mercury vapour lamps or even deuterium lamps were used. Among others, copper deposition could be demonstrated on a substrate at nearly room temperature [2]. Irradiation of palladium acetate layers by Xe_2* light at 172 nm lead to a 0.1μm thick palladium film which was reinforced by an electroless copper plating process. In that way large area structured copper layers could be produced.

[1] B. Eliasson, U. Kogelschatz, Appl. Phys. B46, 299 (1988)

[2] H. Esrom, G. Wahl, U. Kogelschatz, EURO CVD 7, Perpignan (F), paper no. 89, 1989

A COMBINED PULSE CONTROLLED DISCHARGE WITH A POSITIVE V/I CHARACTERISTIC

Valery Godyak
GTE Laboratories Incorporated, Waltham, MA 02254, U.S.A.

It is known that the volt-ampere (V/I) characteristic of a gas discharge has a very small (or, even negative) differential resistance, Rd=dV/dI, that is much less than its static resistance, Rst=V/I. This discharge feature necessitates the use of a ballast for stable discharge operation with a voltage driving source.

The nature of a small (negative or positive) differential resistance in a low pressure discharge is such that the electric field (which maintains the discharge) provides two different functions - ionization and current conduction (accompanied by electron heating). The ionization balance in the discharge requires the electric field to be fixed (as the temperature of boiling water) independently from the ionization (evaporation) rate. In this condition the plasma density is nearly proportional to the discharge current and the differential plasma resistance is close to zero.

The functions of ionization and current conduction can be separated to overcome the effect of small differential resistance. This is achieved in non-self sustained discharges where ionization is provided by some external means (UV-radiation, electron beams, etc). In these discharges the plasma density (and also the conductivity) mainly depends on the intensity of the ionizing source, thus providing an essentially positive (that is, Rd≅Rst) V/I characteristic for the discharge.

A different approach to achieve a non-self sustained discharge with an essentially positive V/I characteristic is discussed in the present report.

The overall concept of this approach is to temporally separate the ionization process and the energy pumping process through a combined pulse-DC or pulse-AC driven discharge. In this approach ionization is fully provided by short over-voltage pulses while energy deposition is supplied by DC or AC current in the afterglow stage between ionization pulses.* Three circumstances permit easy access to this non-self sustained regime under conditions that are typical for low pressure light sources. These are:

1. The ionization power losses are just a small fraction of the discharge power (for fluorescent lamps-just a few percent).

2. The ionization rate is a very sharp function of electric field ($z=E^k$; $k>>1$).

3. The electron energy relaxation time, te, is much smaller than the ambipolar diffusion time, td. (microsecond and millisecond time scales, respectively).

The results of modeling and experimentation on combined pulse-DC and pulse-AC discharges in Ne/Hg and commercial fluorescent lamps filled with Ar/Hg will be given in the presentation. Particularly, the existence condition for such a combined discharge was found to be as follows:

$$\Theta = \frac{tp}{t} = \frac{Te}{Teo}\left(\frac{Eo}{Ep}\right)^k$$

where Θ is the pulse duty factor, Te and Teo are electron temperatures in the afterglow and the self sustained regimes, respectively, and Eo and Ep are the electric fields in the self sustained regime and pulsed regime, respectively. It will be shown that the electron temperature, Te, and the electric field, E, between the ionization pulses depends on the DC or AC discharge current while the plasma density is determined by Ep and the pulse width, tp.
The plasma density in the afterglow stage can be fixed provided that the pulse period t is much smaller than the diffusion time td. Calculations show and our experiments confirm that pulse controlled discharges with an essentially positive V/I characteristic can be achieved with the average pulse power being much smaller than the total discharge power. Pulsed, DC and AC characteristics as well as the light output of the pulse controlled fluorescent lamps will be presented.

An example of the transition from a negative V/I characteristic to a positive one in a fluorescent lamp is given in here for different ionizing pulse intensities.

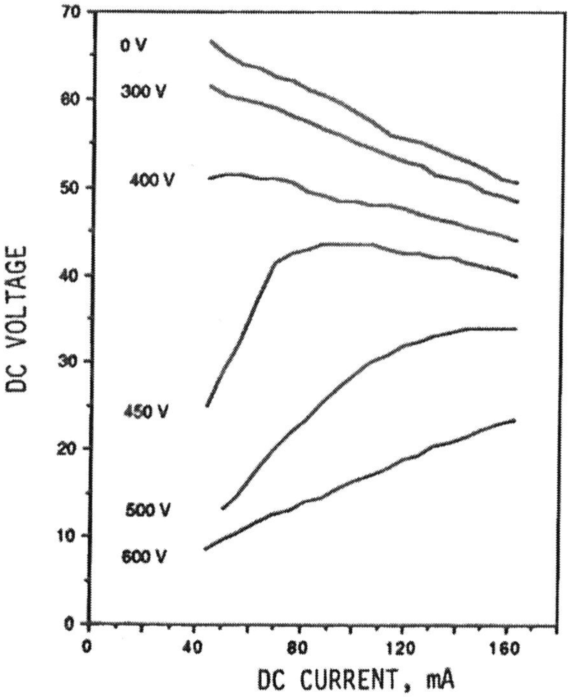

*V. Godyak, US Patent 4,792,727

Radiation characteristic of flash- and arc lamps for laser pumping

P. Greve

Carl Zeiss, L-FLab, 7082 Oberkochen FR Germany

1. Introduction

Most solid state lasers today are still being pumped by high power flash- or arc-lamps filled with xenon or krypton gas. For highly efficient pumping the amount of radiated energy which is concentrated in the laser rod should be as high as possible. Therefore a ray-tracing program has been developed [1] which follows each ray from the surface of the emitting lamp plasma throughout all transmissions and reflections to the rod. For exact calculations it is necessary to know the starting parameters quite well. That's why measurements have been done on the spectral emission, reabsorption and emission indicatrix of several flash- and arc-lamps.

2. Experimental set-up

Spectral emission and absorption are measured with a compact diode-array-spectrometer (Zeiss MCS) described in detail elsewhere [2]. The spectral resolution is about 1 nm, which is quite sufficient for our purpose. For the convenience of the user it can be fiber coupled to the light source. No absolute calibration has been done. The reabsorption has been measured integrally by refocusing the emitted light to the lamp thus doubling the unidirectional detected energy, but for the losses of the arc plasma and the external optical elements [3]. All losses which do not origine from the plasma are distinguished by an extra measurement when the lamp is not operated. Finally the emission indicatrix is evaluated by a fiber detector which is moved angularly around the emitting lamp.

3. Experimental results

The main experimental result is that the arc plasma is not a surface emitter and does not emit with an indicatrix $I \;\; \cos^2$ under each operating conditions. Assuming a model function $I(O) \cos^x O$, the parameter x depends from the bore size, pulslength and current density and varies from about 8 for cw-operation with low current density to 2 for long pulsed operation with high current densities (Table 2). In the cases where the plasma is not optically thick the reabsorption A is <1. For cw-operation the transmission T = 1-A is given in Table 1. With these experimental parameters a modelling within a ray tracing program is possible.

[1] P. Greve et al.; Eureka report 13 EU 017, Juli 1987
[2] H. Schlemmer; Simultanspektrometer, Zeiss Report June 1984
[3] P. Greve et al.; Feinwerktechnik & Meßtechnik 96, No. 6, 245 (1988)

Tab. 1: Total Transmission of Arc plasma

Wavelength	Current density (A/cm^2)			
nm	120	160	200	223
760	0.29±0.11	0.28±0.08	0.23±0.06	0.27±0.04
810	0.25±0.07	0.25±0.05	0.20±0.05	0.19±0.02
828	0.29±0.04	0.34±0.13	0.26±0.06	0.22±0.05
877	0.44±0.07	0.34±0.14	0.24±0.12	0.28±0.06
892	0.36±0.08	0.32±0.09	0.19±0.11	0.24±0.08

Tab. 2: Experimental fit parameter for emission indicatrix

Current density (A/cm^2)	Pulslength (ms)	Exponent X
200	cw	8
445	0.36	6.5
754	0.65	4.3
1580	0.36	4.3
490	2.0	3.6
980	2.0	3.2
3175	0.36	2.9
2380	2.0	2.1
3250	2.0	1.9

PULSED OPERATION OF HIGH PRESSURE SODIUM LAMPS

K. Guenther, H. G. Kloss, T. Lehmann,

Kombinat NARVA Berlin, DDR

R. Radtke, F. Serick,

Zentralinstitut fuer Elektronenphysik, AdW der DDR

High pressure lamps with color temperatures (CCT) of about 3000 K, high efficiency, and an acceptable color rendering index (CRI) would be favourable sources for indoor lighting. The excellent performance during life of high pressure sodium (HPS) lamps in connection with their high efficiency recommend them as a good starting point to reach this object.

It is well known that the CRI can be improved effectively by raising the sodium pressure. Unfortunately, the CCT can be improved only slightly, and the efficiency and the lamp life decrease considerably.

Another attempt was made by feeding the lamps using short energy-rich power pulses instead of the customary AC supply. The result was a considerable increase of the CCT connected with a lowering of the light output [1].

We repeated these investigations because the drop of efficiency is not a necessary result of the pulsed operation mode: It is obvious that a high instantaneous power increases the plasma temperature. The result is a proportional increase of heat conduction losses and an exponential increase of radiation. Thus, in contrast to the results [1] an increase of the radiation efficiency should be expected.

To explain the results of [1] we estimated the plasma conditions during a high power pulse for a standard 70 W HPS lamp with p_{Na} = 16 kPa and p_{Hg} = 82 kPa. Using a parabolic (n=3) temperature profil the current balance equation yields temperatures until 7000 K in the axis of the discharge for the experimental data of the electrical conductivity at pulse currents up to 20 A. For such temperatures the number density of neutral sodium atoms decreases considerably compared with the stationary conditions at about 4200 K, and the "Normtemperatur" lying for the most efficient sodium resonance lines at about 6000 K is exceeded. Therefore, the sodium radiation decreases in favour of the radiation of mercury. Due to the lower efficiency of the mercury radiation in comparison with the sodium radiation the total yield of the discharge becomes smaller.

In contrast, calculations of the same kind executed for pure sodium with p_{Na} = 36 kPa and for sodium-rich Na-Hg-plasmas with p_{Na} = 24 kPa and p_{Hg} = 30 kPa show a slighter increase of the plasma temperature at the same pulse current, confirming our expectation of a better efficiency at pulsed operation.

Experiments using various molar ratios Na/Hg and various pulse parameters - instant power, pulse shape and frequency - have been performed. By means of complex optimization we obtain HPS lamps with a color temperature of 3000 K and a luminous flux of 5500 lm at 70 W power input.

[1] Brates, N. and Wyner, E.F., J. Illum. Engn. Soc., New York, 16 (1987) 2, p. 50 - 66.

OPERATION OF LOW WATTAGE ELECTRODELESS HID LAMPS

D. Hollister, M. Neiger† and S. Berman

Applied Science Division
Lawrence Berkeley Laboratory
University of California
Berkeley, California 94720 U.S.A.

Abstract

Electrodeless HID lamps offer a number of advantages over electroded ones. Lamp ingredients can be selected which would react adversely with electrode materials in electroded lamps, thus offering new possibilities for luminous efficacy and spectral tailoring. Due to the absence of cold spots behind electrodes higher vapor pressures of metal halides seem possible with the prospect of higher luminous efficacy, or more stable ingredients (at wall temperature) can be selected at similar vapor pressures offering the prospect of longer lamp life. Lamp bulb manufacture seems greatly simplified in the case of electrodeless lamps for example, spherical or pill box shaped bulbs. This is especially interesting for low wattage lamps below the 100 watt level, where reproducible bulb manufacture poses problems in the case of electroded lamps.

† Visiting researcher at LBL on sabbatical from the Universität (TH) Karlsruhe, Lichttechnisches Institut, Kaiserstrasse 12, D-7500 Karlsruhe 1, West Germany

Our electrodeless HID lamps were operated by the electric field induced by the rf magnetic field inside a solonoid. Lamp bulbs were either spherical or pill box shaped with fills of Hg, Na and metal halides. Argon was used as a starting gas. After lamp warm-up a donut shaped toroidal discharge is formed. The discharge clearly is detached from the bulb's wall and a highly luminous toroidal current carrying core surrounded by an outer mantle emitting molecular radiation is easily identified. The appearance is that of an electroded plasma column bent into a closed ring.

The operating frequency of our lamps is 27.12 Mhz, lamp major diameter is one-inch or smaller, operating lamp power levels are typically at or below 100 watts with arc loadings in excess of 25 W/cm. Figure 1 shows a typical lamp spectrum at approximately 100 watts of electrical power in a one-inch inner diameter spherical lamp filled with Na/Sc amalgam, ScI_3 and NaI at an operating pressure below 1 atm. Results of spectral lineshape (pressure) and integrating sphere (luminous efficacy) measurements are presented.

This work was supported by the Assistant Secretary for Conservation and Renewable Energy, Office of Buildings and Community Systems, Building Equipment Division of the U.S. Department of Energy under Contract No. DE-ACO3-76SFOOO98.

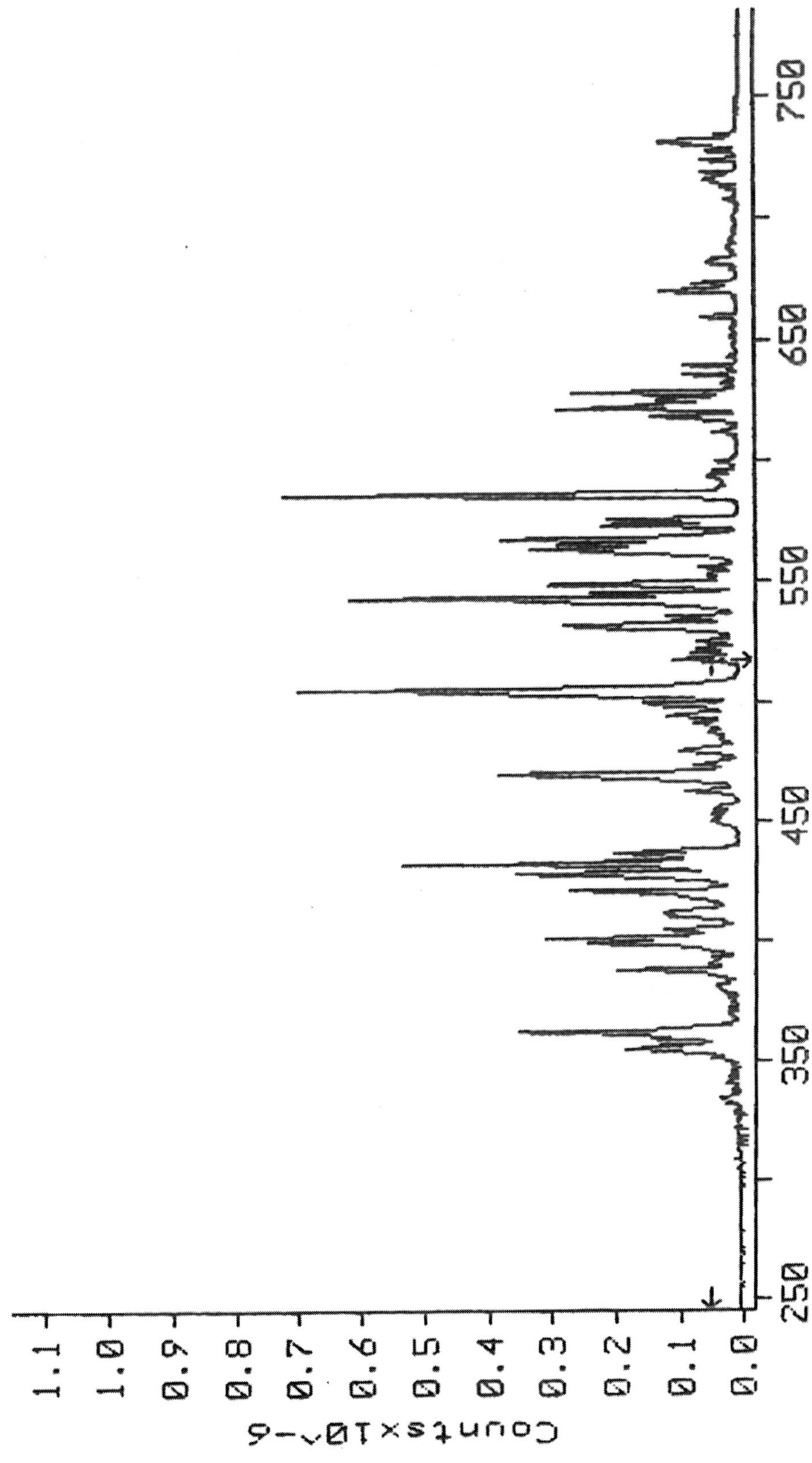
Counts x10^-6
0.0
0.1
0.2
0.3
0.4
0.5
0.6
0.7
0.8
0.9
1.0
1.1
250
350
450
550
650
750

High Power Mikrowave Lamp for Pumping a Nd-YAG Laser*

G. Löhmann, H.-P. Popp, M. Neiger
Lichttechnisches Institut der Universität Karlsruhe, Germany

1. Problem:

Xenon or Krypton lamps normaly are used for pumping the Neodym rod in laser systems, whitch are normaly used in industry for material processing and preparation. Such laser systems usually use DC lamps, which only have a short life and pose problems in replacement. The most important reason for the short life of DC-lamps is, that small particles of the cathode material are deposited on the inner quartz surface or migrate into the quartz which causes themal stress and later subsequently cracks. An explosive failure of the lamp is often the result.

2. Solution:

In the Lichttechnischem Institut we have developed a elektrodeless microwave powered lamp. The microwave is generated by an industrial magnetron . The frequenzy used is 2.45 GHz.
The lamp is very easy to produce and can be simply exchanged even when mounted within a laser system. Filled with Krypton the lamp has a cold fill pressure of typical 1 bar.
For cooling the lamp, a special kind of oil is used , which does not absorb microwave ($\tan \delta < 10^{-3}$), and it is transparency in the optical wavelength range of the absorption band of the laser rod (transmission > 95%). The gas-discharge-lamp is cooled in the cavity itself.

* Supported by BMFT Germany.

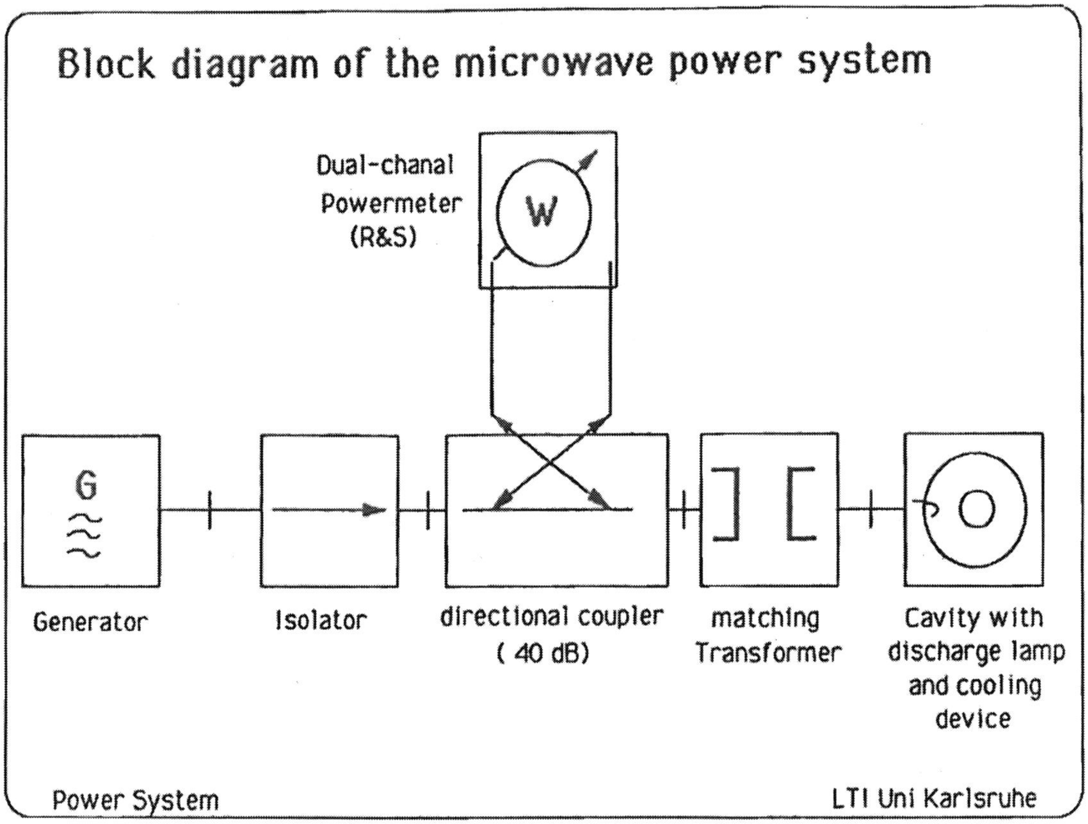
Block diagram of the microwave power system
Dual-chanal
Powermeter
(R&S)
W
G
Generator
Isolator
directional coupler
(40 dB)
matching
Transformer
Cavity with
discharge lamp
and cooling
device
Power System
LTI Uni Karlsruhe

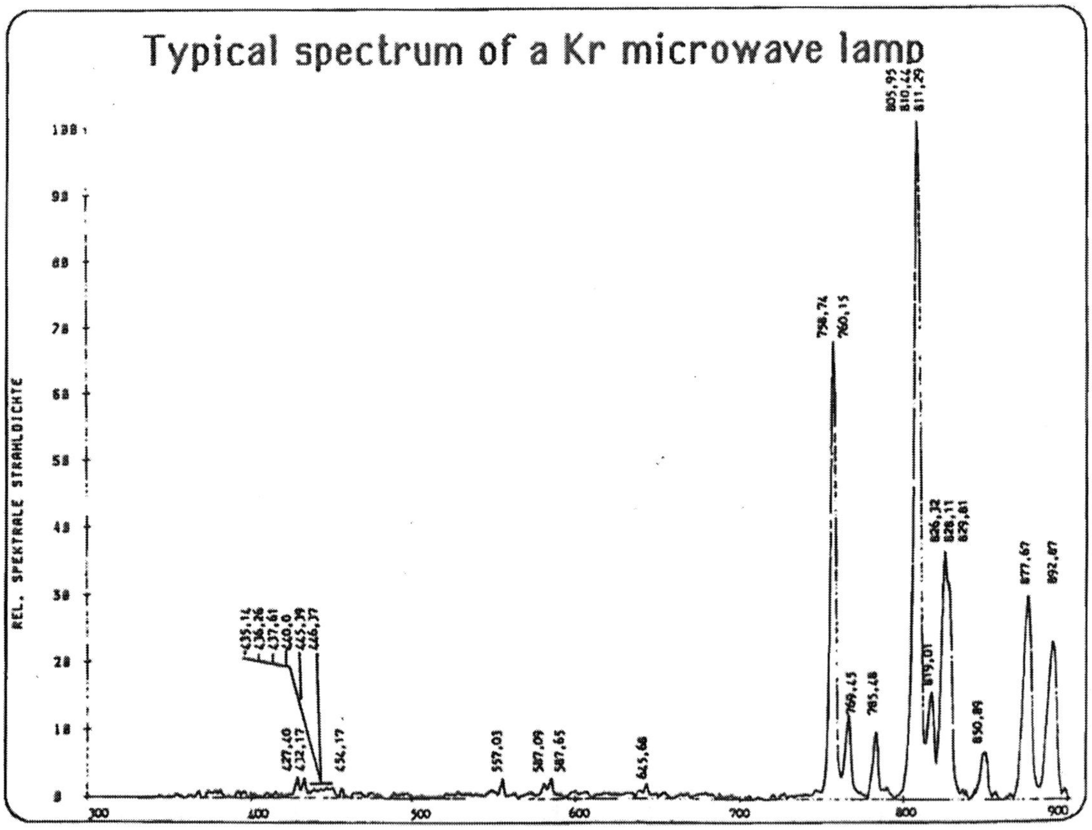
Typical spectrum of a Kr microwave lamp
REL. SPEKTRALE STRAHLDICHTE
300
400
500
600
700
800
900
427,40
432,17
435,14
436,26
437,61
440,0
445,39
446,37
456,17
557,03
587,09
587,65
645,66
758,74
760,15
769,45
785,48
805,95
810,44
811,29
819,01
826,32
828,11
829,81
850,89
877,67
892,87

Novel Circuits for High Frequency Operation of HID Lamps

E.J.P. Mascarenhas and A.L.G. Edwards
THORN EMI Central Research Labs, Dawley Road, Hayes, Middx., UK, UB3 1HH.

1. Introduction

High frequency operation of High Intensity Discharge lamps has a number of advantages over conventional gear: reduced size and weight, flicker free light, improved luminous efficacy through life, fast run-up and hot re-strike capability (1). However, to operate HID lamps successfully above audible frequencies, acoustic resonance has to be overcome (2). A number of techniques to achieve this have been suggested, the most common of which are commutated DC (3), square wave current (4), very high frequency operation (4), and frequency hopping (5).

The first two techniques require a full bridge circuit topology and the second two a half bridge; furthermore, each technique places different limitations on the lamp voltage and current waveforms and specific requirements on the electronic circuit.These aspects of an electronic ballast together with economic and circuitry limitations have been investigated and a novel frequency hopping system is also presented.

2. High Frequency Operation

Both square wave current and commutated DC techniques require lamp currents with very low levels of AC modulation to prevent resonance . Experimental results for a 150W 3000K metal halide lamp (Arcstream - HQI-T type lamp) show that the modulation has to be below 5% for no observable acoustic resonance, hence relatively large inductors are needed. Additionally, the rise time of the square wave has to be less than 5 µs. Running a HID lamp at frequencies greater than its acoustic instability range will result in stable lamp operation. At 150kHz most lamps operated satisfactorily, however some needed to be operated at frequencies greater than 300kHz. Arc bowing, similar to that at 50 Hz, was observed for these three techniques.

3. Frequency Hopping Ballast

Efficiency and cost considerations point to a half-bridge system operating in a frequency range of 15-30kHz as being the optimum system for an electronic ballast. In this way the number of power components, the switching losses and RFI problems are minimised. In terms of acoustic resonance this is the most difficult area to design such a ballast but frequency hopping is one technique that can be used.

Successful operating on a frequency hopping system requires that acoustic resonance free windows exist. For the Arcstream lamp an acoustic free window approximately 2 kHz wide occurs in the range 19 to 23kHz. It was not possible to operate the lamp on a single frequency in this window, however stable operation was obtained by frequency hopping using a novel frequency hopping algorithm.

A programmable ballast has been designed and built to evaluate the arc stability of a 150W metal halide ARCSTREAM lamp operated using various frequency hopping algorithms. It was found that the only algorithm which could operate a large number of lamps stably used a rotating diameter sequence. No feedback is needed to maintain stability which significantly reduces the need for complex electronics, and the entire control system can be implemented using a single digital ASIC. A stable straight arc was observed with a small decrease of the CCT compared to 50 Hz operation. Preliminary life test data indicates that improvements in the maintenance of the luminos flux through lift are obtained compared to 50Hz.

4. References

1) H. Stromberg & R. Schafer, Lighting Research Tech. Vol.15 No. 3 1983
2) H. Witting, J. Appl. Phys. Vol. 49 No. 4 May 1978
3) H. Faehnrich & E. Rasch, J. Ill. Eng. Soc. Summer 1988
4) Koshimura et al, 20th CIE, Sept. 1983
5) H. Lange, 3rd. Int. Sym. Science & Tech. of Light Sources. 1983

Efficient Production of Incoherent Excimer Radiation in Dielectric Barrier Discharges

H. Müller, M. Neiger, V. Schorpp and K. Stockwald

Lichttechnisches Institut, University of Karlsruhe, D–7500 Karlsruhe, FRG

Dielectric barrier discharges in high pressure gases are transient, nonequilibrium glow discharge plasmas capable of producing selective, incoherent radiation with good efficiency. We investigated incoherent excimer emission from such discharges, both experimentally and by a simple theoretical modelling approach.

Fig. 1 shows a schematic of our discharge device, which is electrically fed by 1 – 10 kV sinusoidal voltage at frequencies ranging from 50 Hz to 300 kHz. Gas fillings used in this investigation were rare gas /Xe/HCl and rare gas/Hg mixtures at total fill pressures in the range 0,1 – 2 bar. Fig. 2 shows typical oscilloscope signals for a Ar/Xe/He–HCl–discharge: the sinusoidal feed voltage and discharge current spikes (Fig. 2a), the transient current signal of a individual microdischarge (Fig. 2b), and the XeCl 308 nm excimer emission pulse of a single microdischarge. Fig. 3 shows temporally and spatially averaged emission spectra: $XeCl^*$ (308 nm) from a Ar/Xe/He–HCl discharge (Fig. 3a), ArF^* (193 nm) from a Ar/SF_6 filling (Fig. 3b), Hg (185, 254 nm) and $XeHg^*$ (210, 270 nm) from a Xe/Hg–discharge (Fig. 3c). Quantitative measurements of average electrical input power and optical output power in the e. g. $XeCl^*$ excimer band yielded efficiencies up to 10 %.

Because of the large number of possible discharge parameters (gas–mix, partial and total pressures, gap distance, dielectric thickness and permeability, operating voltage and frequency) a theoretical description of a single microdischarge is highly desirable. A first one–dimensional modelling approach, which describes both the electrical parameters of the barrier discharge and the nonequilibrium discharge chemistry yields already quite satisfactory agreement with our measured data.

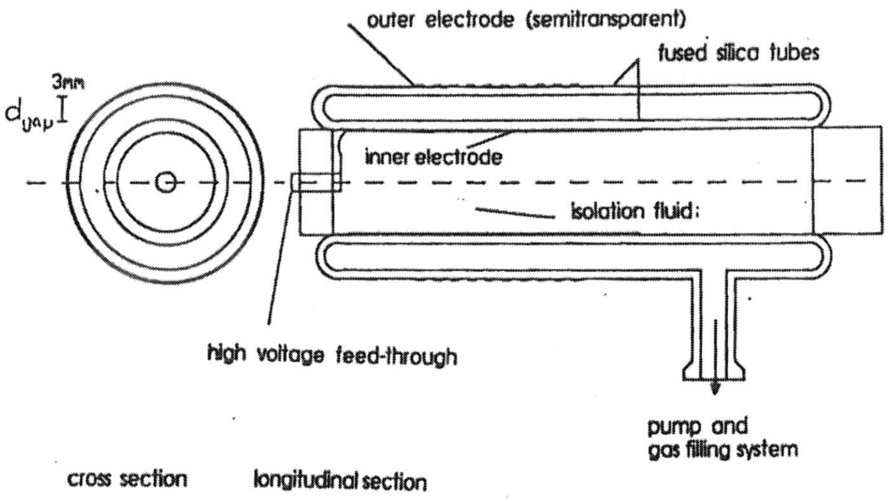

Fig.1: Schematic of discharge device

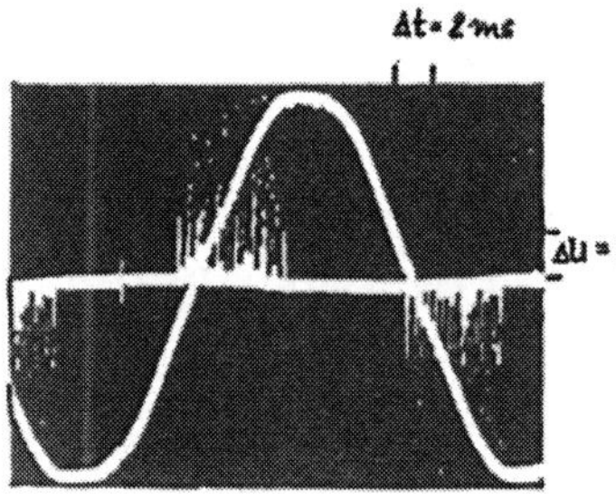

Fig. 2 a: feed voltage and correlated current spikes

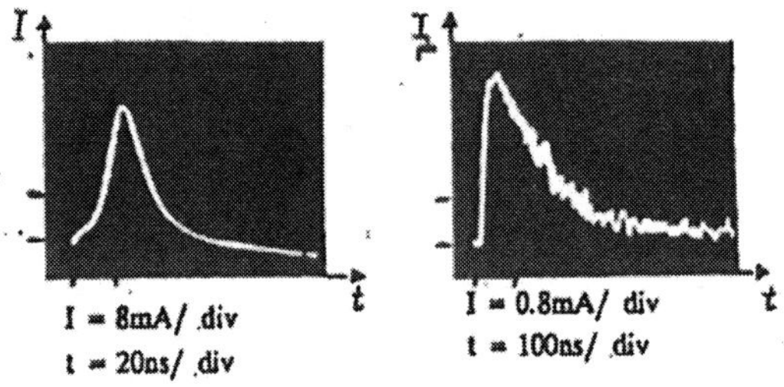

Fig. 2 b: translent electrical current signal and correlated emission pulse

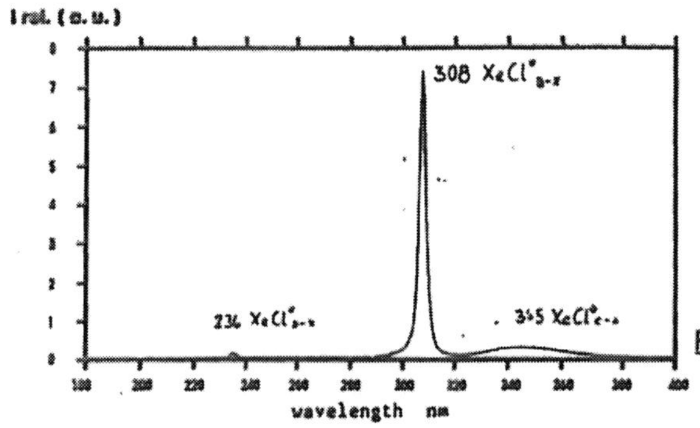

Fig. 3 a: XeCl *-spectrum 308 nm

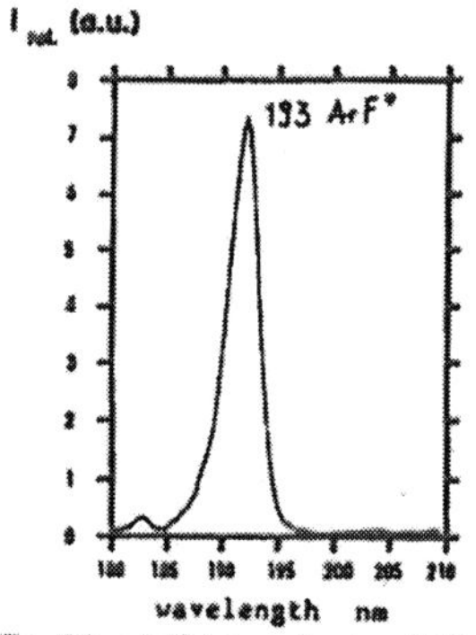

Fig. 3 b: ArF *-spectrum 193 nm

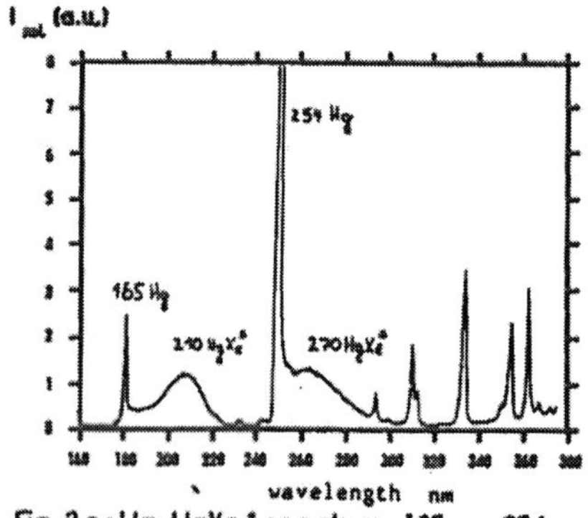

Fig. 3 c: Hg. HgXe *-spectrum 185nm, 254 nm; 210nm, 270 nm

HIGH POWER ARC LAMPS OPERATING IN THE MIDDLE QUENCY RANGE FROM 10–100 KHZ

eichert[**], K.–J. Dietz[*], A. Dohn[*], U. Krönert[*], G.Mayer[**], M. Neiger[**], Popp[**]

C. Heraeus GmbH, D–6450 Hanau

chttechnisches Institut, Universität Karlsruhe

ypton and Xenon lamps are commonly used as light sources for pumping d–state lasers. The short lifetime and difficult replacement of the lamps ent serious problems in the application of lasers in industrial material cessing. The limiting factor of lifetime is the evaporation of electrode–material d buildup of light–absorbing deposits within the lamp envelope. This leads to a duction of the light output and at high power levels to overheating and bsequent breaking of the quartz tube.

the lamp is operated by a middle frequency AC–supply instead of standard C–operation, in can be anticipated that the lifetime will be doubled according the symmetric distribution of the electrode stress. As an alternative to the rth–alkaline–doped cathode–material which shortens lamp life by enforcing ecristallization of the quartz tube we used thoriated tungsten electrodes. The deposition of evaporated electrode material is reduced by using a tungsten halogene cycle. This requires a special molybdenium cup cooling which provides a sufficient temperature in the dead volume of the tube.

If operated at high frequencies, the lamp is subject to arc instability due to acoustic resonances. The resonance frequencies depend on the geometry of the arc tube interior and the velocity of sound in the discharge /1//2/. The optimum frequency windows have been investigated in the frequency range up to 80 kHz.

As shown in Fig. 1 the radiative output is independent of the operating frequency. Though the modulation amplitude of the radiation decreases with increasing power and frequency. Due to the mode of operation reported here, the efficiency of the lamp is increased in comparison to standard DC–operation (see Fig. 2).

This work was founded by the German Federal Minister of Research (BMFT).

/1/ Witting, H. L.
Journal of Applied Physics, May 1978, Vol. 49

/2/ Statnic, E.
Technisch wissenschaftliche Abhandlungen der Osram–Gesellschaft
1986, Bd. 12, Seite 394.

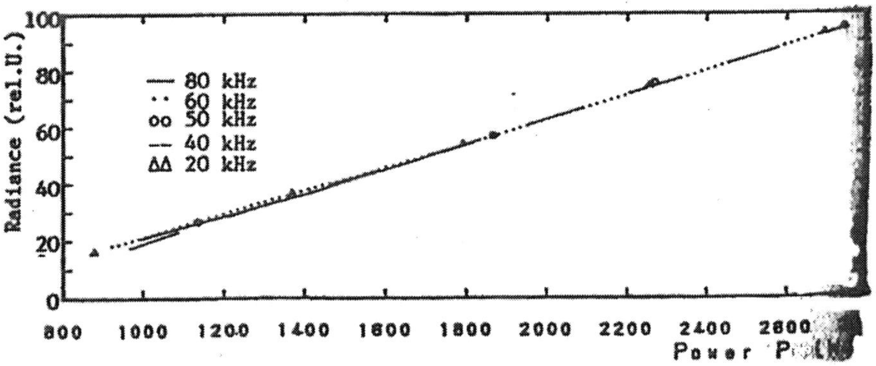

Figure 1

Radiative lamp output as a function of electrical input power for different operation frequencies in the range 20–80 kHz

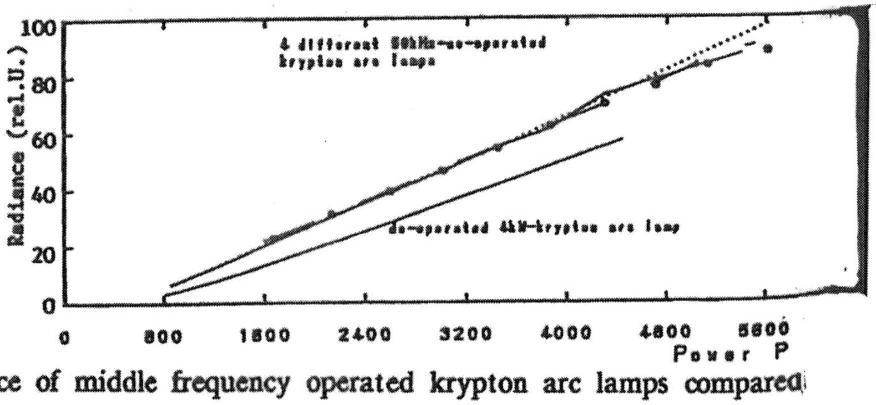

Figure 2

The radiance of middle frequency operated krypton arc lamps compared DC–operated arc lamp of same dimensions.

Circuitous Model of gas Discharge Lamp.

Wang Lian-cai, Yang li-xi,

Qin Ke-jing, Jia Geng-xin.

Zheng Zhou Institute of Technology, China.

Abstract:

The circuitous model of gas discharge lamp in the working condition always are regarded as a resistance and inductive circuit (). But according to the voltage wave and the current wave of gas discharge lamp in the working condition, and taking the gas discharge lamp as an element of energy change which changed the electric energy into light energy and heat energy , all of these have explained that gas discharge lamp is the only nonlinear resistance (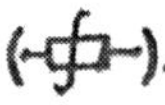).

According to the circuit theory , this paper explains this problem. Having correct understanding of the circuit model will further guide the manufactory of the lamp, the design and the production of ballasts, and there will be a series of new calculating methods to the circuit of the gas discharge lamp.

Electronic Ballast for High Pressure Sodium Gas Discharge Lamp

Armin Wegener and Theo Hüsgen

Philips GmbH, Research Laboratory Aachen, Postfach 1980, D-5100 Aachen, W.-Germany

Strong efforts are taken by many companies to improve the performance of a motorcar headlight system. The improvements can be achieved by using a different reflector geometry /1/ or by applying new lamps as light sources such as high pressure gas discharge lamps. Besides the commonly used Metal Halide Lamps which were optimized for this application /2/, /3/, /4/ there are also high pressure sodium gas discharge lamps (HPS) which are suitable for automotive lighting /5/. A HPS lamp-ballast system was investigated to evaluate its advantages and drawbacks.

In this paper an electronic ballast for a HPS-lamp driven from a 12 V car battery is presented. The lamp is designed for a motorcar headlight system and shows significantly better light properties compared with the conventional halogen lamp. Although the efficacy of the HPS-lamp is 40 to 60% lower than that of a Metal Halide lamp, it is still a factor 2 higher compared with a halogen lamp allowing therefore lower wattage. Gas discharge lamps have a longer lifetime and due to their higher luminance over halogen lamps the beam performance and intensity can be considerably increased, so that more compact headlight units can be realized. The luminance of the HPS-lamp ranges between that of a Metal Halide and a halogen lamp.

The concept of a ballast depends on the characteristics of the gas discharge lamp as well as on the requirements for the whole ballast-lamp system. The ballast has to provide a pure ac-lamp current, because the HPS-lamp cannot be operated with a dc-current. Even a dc-component must be avoided, because otherwise a strong corrosion of the burner material can be expected. The possible frequency range of the lamp current is restricted by acoustic instabilities of the arc which occur at frequencies of appr. 10 kHz up to 200kHz. So a square wave lamp current was chosen with a frequency of about 100 Hz generated by an inverter (Fig.1). Owing to the negative V / I characteristic of the gas discharge the lamp current has to be stabilized within this l.f. period by h.f. switching, causing an h.f. ripple superimposed on the l.f. square wave (Fig.3).

The input voltage is stepped up by a boost converter stage to the lamp voltage during steady state operation and up to an open circuit voltage of 180V during ignition of the lamp. Superposed on this open circuit voltage a transformer generates high voltage pulses of a few Kilovolt to ignite the lamp (Fig.1).

It is very important to have "instant" light output after switch-on (cold start). Therefore the lamp is supplied with the maximum power level it can tolerate for a short time. The light output and the power delivered to the lamp are shown in Figs. 4a,b.

This electronic ballast provides fast run-up and hot restrike for the lamp and includes various control and protection features, which can be recognized in Fig.2. They are colour temperature control, minimum and maximum power levels, lamp current control and a time, lamp state and battery voltage dependent power control after switch-on. Thus a very reliable and stable operation of the HPS lamp at high efficiency ($\eta > 82\%$) is achieved. But due to the above stated advantages of Metal Halide lamps over HPS-lamps it seems that a headlight system with a Metal Halide lamp is more favourable.

References:

/1/ G. Lindae: Verbesserung des Automobil-Abblendlichtes durch Scheinwerfer mit polyelliptischem System; Bosch Technische Berichte, Vol.8, No.3, pp 107 (1986)

/2/ Motorcar Headlight System MHS-35W;
Brochure, Philips Central Development Lighting Division, 1985

/3/ anonym: Discharge lamps for cars 'by the 1990s';
Electrical Review, Vol.216, No.5, 8.Feb. 1985, p.20

/4/ J.M. Davenport, R.L. Hansler: Instant Light for Metal Halide Lamps;

Illuminating Engineering Society, Conference, Minneapolis, Aug. 1988

/5/ G. Szekacs: Low wattage HPS lamps for vehicle headlamps; Illuminating Engineering Society, Journal, Summer 1986

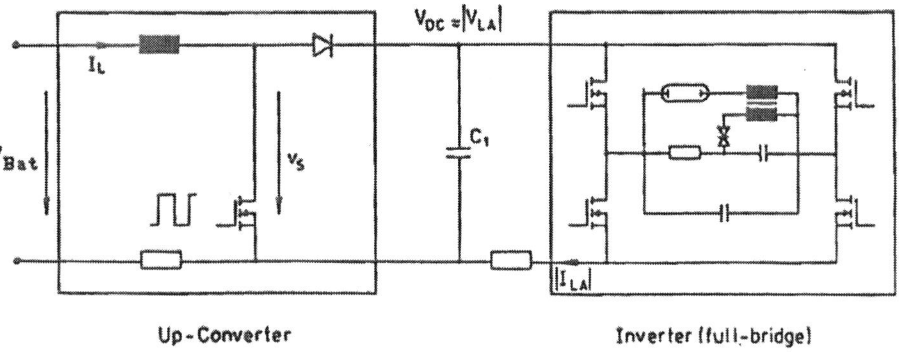

Fig. 1 Block diagram of the ballast for HPS-Lamp

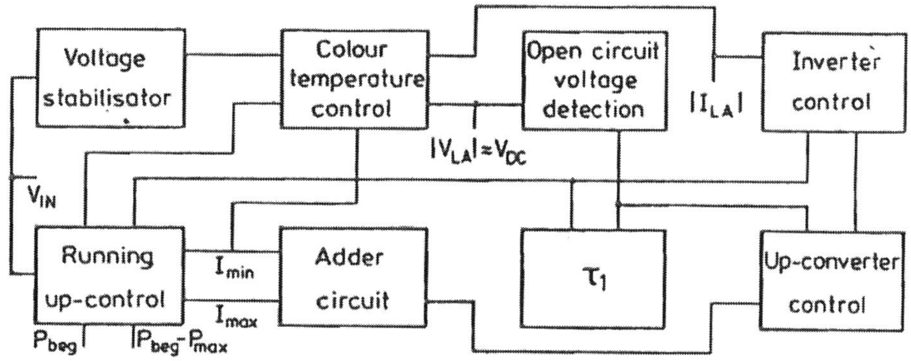

Fig. 2 Block diagram of control circuit

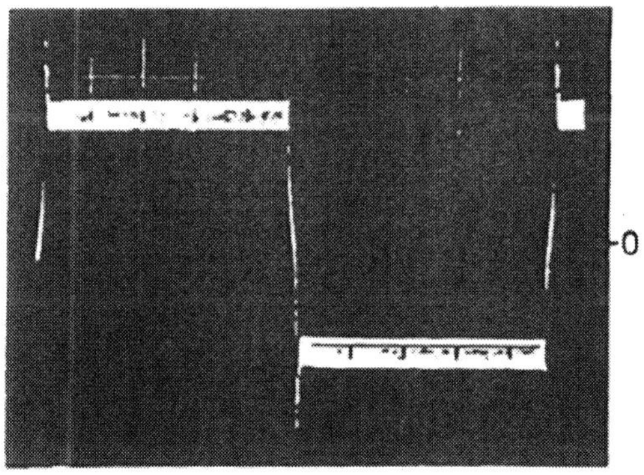

Fig. 3 Lamp current [.5A/Div], t=1ms/Div

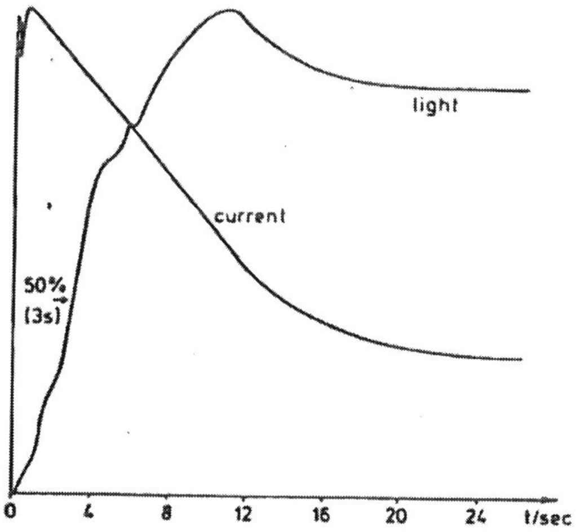

Fig. 4a Lamp current and light output after switch-on

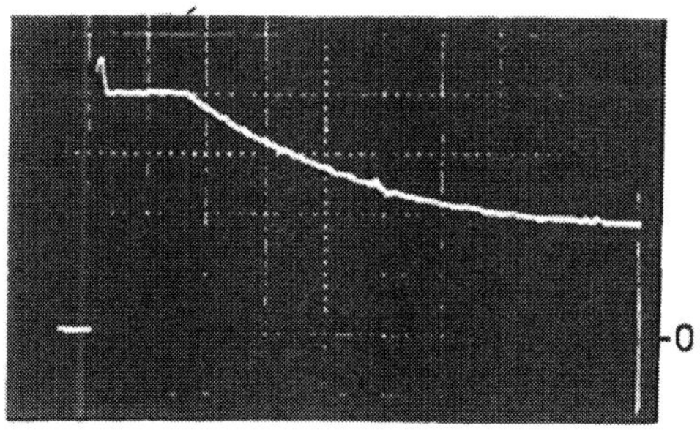

Fig. 4b
Lamp power after switch-on [20W/Div], t=2s/Div

Simplification of Electronic Ballast for Self-Ballasted Fluorescent Lamps

N.Yoshikawa and T.Imai

Light R&D Center, Matsushita Electronics Corp.

Takatsuki, Osaka, Japan

Conventional electronic circuits for self-ballasted compact fluorescent lamps employ rather complicated design. With the amalgam-filled lamp, simplification of the circuit has been realized through introduction of a specific one-transistor type inverter.

The new circuit features its simplified design ; one-transistor blocking type inverter with only one choke coil L_1' against two choke coils L_1 and L_2 in conventional types as shown in Fig. 1 (a) and (b). Furthermore, this offers a reduction in switching loss of the transistor Q because of its lower collector peak current, thereby leading to higher inversion efficiency above 90 %.

In the above design simplification, it is noted that the circuit brings about asymmetrical current wave form containing d.c. component as shown in Fig. 2. Here, a technical significance is that cataphoresis phenomenon due to the d.c. current component has been prevented by setting the amalgam in the lamp at its anode side relative to the d.c. current component (Fig. 3). It can be considered that migration of mercury ions to the cathode side by cataphoresis phenomenon is compensated by diffusion of mercury atoms toward the amalgam-set anode side.

Using the above circuit together with the double-folded discharge tube, a 17 W self-ballasted compact fluorescent lamp has been developed. The lamp offers relatively high system efficacy of 65 lm/W.

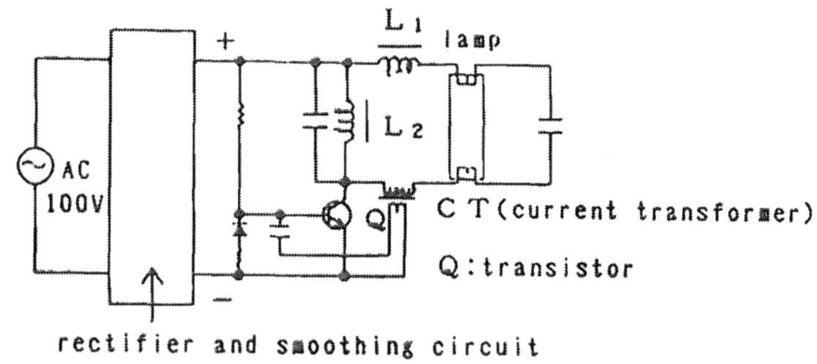

(a) Conventional circuit

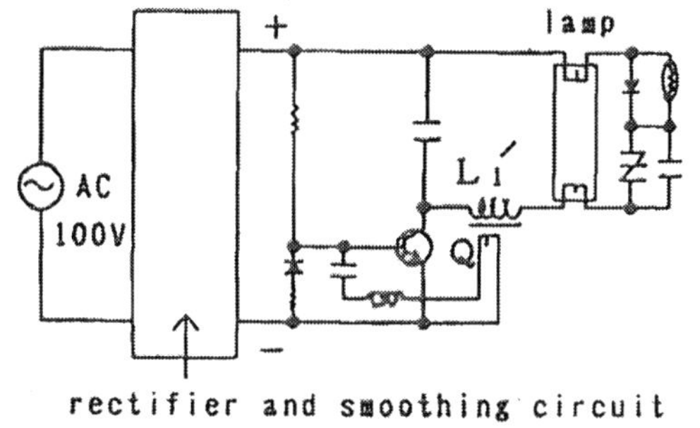

(b) Simplified circuit

Fig. 1 Simplification of electronic circuits

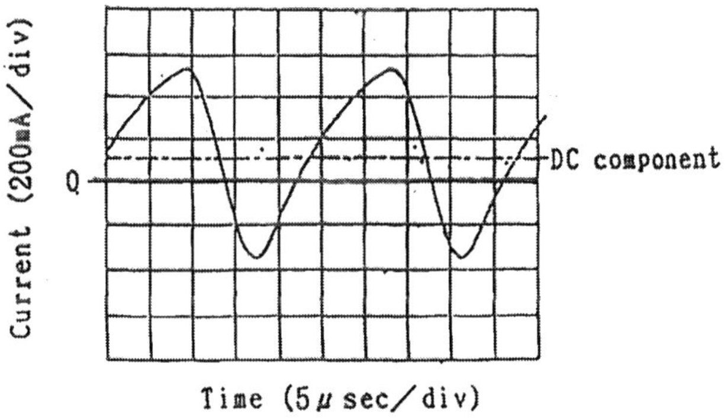

Fig. 2 Discharge current of simplified circuit

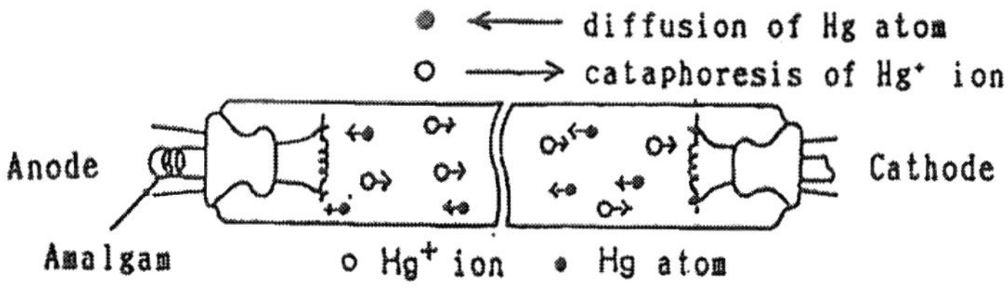

Fig. 3 Principle of prevention of Hg cataphoresis phenomenon by amalgam

UV Radiation from Hg Resonance Lines by Means of Dielectric Barrier Discharges

B. Gellert, B. Eliasson, U. Kogelschatz, Asea Brown Boveri Corporate Research, 5405 Baden / Switzerland

Dielectric barrier discharges were investigated experimentally as well as theoretically. The discharge was ignited in an annular discharge gap between coaxial quartz cylinders (Fig. 1).

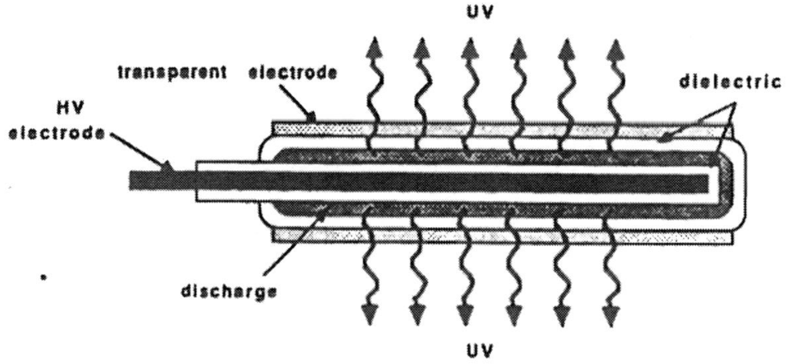

Fig. 1

Many microdischarges are formed when AC fields with frequencies ranging from 50Hz to several hundred kHz were applied. It is demonstrated that it is possible to influence the electron energy distribution in the dielectric barrier discharge by external means.

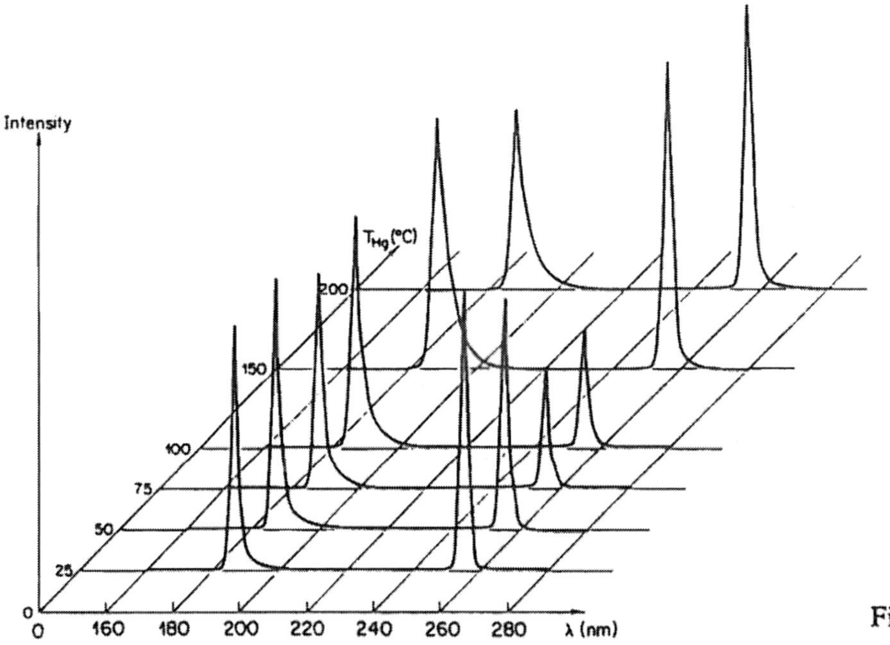

Fig. 2

To investigate the influence of the electron temperature we have chosen a mixture of Hg with various rare gases [1]. Varying Hg vapour pressure by heating the source from 25°C to

200°C results in spectral changes shown in Fig. 2 using 850 mbar Ar and 2 mbar He as a buffer gas. More than 3 times higher intensity can be obtained at 185 nm compared to the emission at 253.6 nm. Absolute intensities were measured for 8 Hg lines in the wavelength region 260-550 nm. Corresponding excited state densities follow closely an exponential dependence on excitation energy. Thus a Boltzmann temperature could be derived.

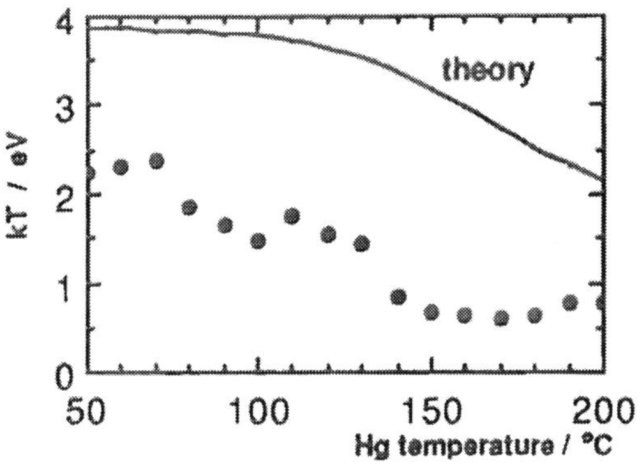

Fig. 3

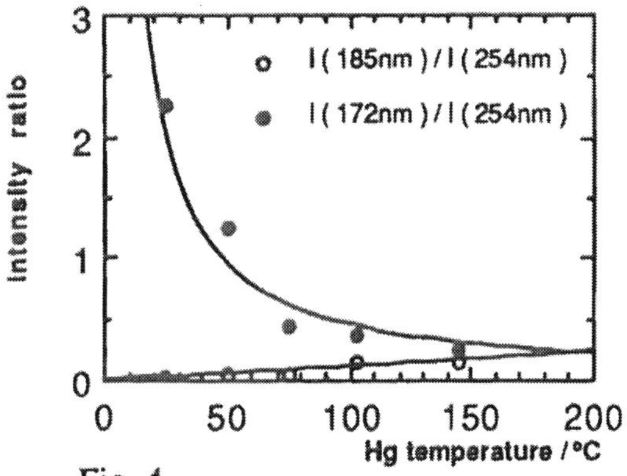

Fig. 4

This temperature shows the same dependence on Hg vapour temperature as does the electron temperature derived theoretically (Fig. 3). The model calculations are based on the solution of the Boltzmann equation taking the reduced Paschen breakdown field for Ar into account. Using a gas with excitation levels close to the resonance levels of Hg, e.g. Xe, wavelength tuning could be achieved by varying the temperature of the UV source. At low temperatures 172 nm radiation of Xe_2* is dominant. Rising temperature increases the emission of the Hg resonance lines. Above 50°C either 185 nm or 253.6 nm radiation dominates depending on Hg temperature (Fig. 4). Using transparent inner electrodes sources radiating inwards can be built. Also flat panels are possible (Fig. 5). The new type of UV-source is thus an ideal tool for many applications in photochemistry and photophysics.

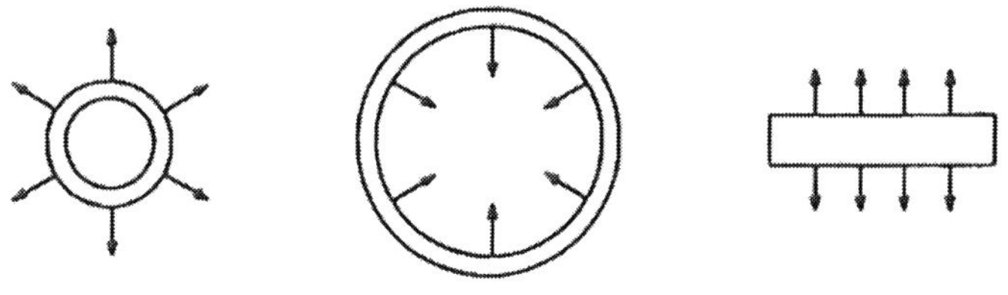
Fig. 5

[1] B. Eliasson, B. Gellert, U. Kogelschatz, Proc. 41st GEC, Minneapolis, USA, 1988

ELECTROMAGNETIC COMPATIBILITY OF LIGHTING SYSTEMS.

J.R.Bergervoet.

Philips Lighting B.V. Eindhoven, The Netherlands

This paper deals with the unwanted high frequency signals, that can be produced by electric lighting systems. These signals can escape as electromagnetic radiation, but they can also be conducted via the connections with the power grid and between different parts of the system, such as light sources, (electronic) ballasts, control units, light sensors, etc. Electromagnetic radiation is always produced by electric currents. High frequency currents between different parts of the system can be an important source of radiation. Therefore it is important to look at complete systems.

A convenient way to describe the radiation is to divide it into electric dipole and magnetic dipole radiation (fig.1). It will be shown that both can be important for lighting systems.

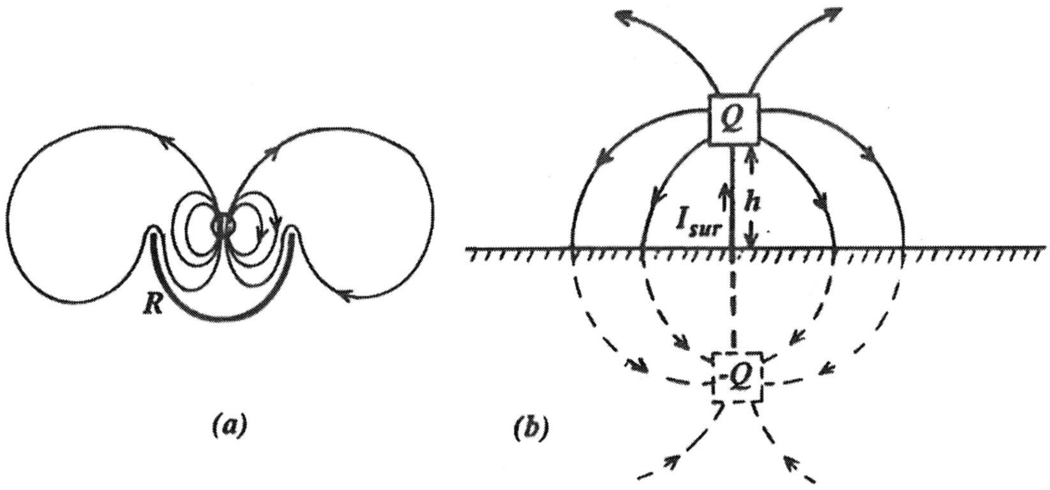

Fig.1 (a) A magnetic dipole, shielded by a reflector (R).
(b) Electric dipole and its accompanying current I_{sur}.

An example is given of a simple system, consisting of a lamp and an electronic ballast. The electric dipole radiation in that case is shown to depend on the length of the connection with the power grid, so it is not a function of the system parts alone. This is of special importance for pole-top light sources (street lighting). General methods are given to predict the different types of radiation.

To prevent excessive electromagnetic radiation, shielding of the system parts has to be used. Because signals in the interconnecting wiring are important too, one also uses filters inside the electronic gear, to prevent the propagation of these signals.
The construction of effective filters, especially at higher frequencies, gives rise to some special problems. In particular there are requirements for the mechanical construction, which are briefly described.

Special attention is given to the role of the lamp itself. It can contribute to both the electric and the magnetic dipole radiation. The difference with other parts of the system is that the lamp cannot be shielded at will, since the light should be able to emerge.
However, a special way of shielding is possible, by taking advantage of the optical system. If there is, for instance, a metal reflector around the lamp, than this can lead to a considerable suppression of the radiation (fig.1a).
Calculations are presented, which allow a prediction of the electromagnetic shielding properties for optical systems with different geometries. The agreement with measurements appears to be very good.

For the measurement of magnetic dipole radiation a new method was used, using a large-loop antenna. This method, which was developed at the Philips Lighting laboratories, has also been proposed within the CISPR/A committee. It offers high sensitivity and is hardly influenced by its surroundings, which makes it convenient to use on any location.
A practical method is also given for the measurement of the electric dipole.

PROGRESS IN ELECTRONICS FOR LAMPS AND ITS FORECAST

E. Bolte

Philips GmbH Forschungslaboratorium Aachen, D-5100 Aachen, FRG

Abstract Electronic ballasts were originally introduced merely to replace the magnetic ballasts of the widely used fluorescent lamps. The objective was to save energy by an increase in luminous efficacy, preferably combined with reduced weight, volume and price of the ballast. Employing ultrasonic frequency operation the first goals could be achieved. Progress in control and power semiconductor devices allow not only simply a continued frequency increase and related reduction in volume of storage elements, but also the introduction of several, electronically controlled operational modes comes within the reach of mass produced lamps. Critical aspects of system architecture, circuit topologies and control strategies of electronic ballasts for the relevant lamp types will be discussed.

The electronic ballast as a switched mode power supply (SMPS), however, conducts harmonic currents into the mains and radiates radio frequencies, both of which interfere with remote-control equipment or radio reception. Additionally the lamp operated with hf emits rf. Principles and measures are illustrated how the functional requirements of the total lamp system can be met simultaneously and economically with severe regulations concerning electromagnetic compatibility.

Whether an electronic ballast can combine comprehensive performance and price competitiveness depends a great deal on the availability of reliable and cheap components. A price drop in discrete power devices made electronics attractive for mass products. The impact of power integrated circuits (presently challenging technologists and stimulating the fantasy of circuit designers [14]) on electronic ballasts will be outlined, taking into account availability and physical device constraints. A consideration of the passive, especially magnetic components[13] (not subject to minimaturization via integration) under high frequency large signal operation will finish the review.

I. Introduction

A device converting electrical energy into visible light is referred to as a light source or lamp. But just for highly efficient lamps the electrical energy is by no means available in a form adequate for the lamp. The task of matching the electrical source to the needs of the lamp is performed by an electrical circuit, here referred to as ballast. More specifically the review will focus on circuits employing mainly electronic means. These are the "Electronic Ballasts". For the historical development see Ref [3].

Consequently the electronic ballast is an interface between the power source and the lamp. As an example fig. 1 shows as source the public power system, for mobile applications it has to be substituted by the relevant supply. Basic functions are:

a) for discharge lamps: ignition, re-ignition, starting-up, stabilization of the lamp current, generation of ac supply from dc rail for transport application;

b) for incandescent and tungsten halogen lamps: adaptation of lamp voltage.

Electronic ballasts have been developed for existing lamp systems to improve the properties, initially above all the energy utilization. This is closely related to a comprehensive effort of the lighting industry resulting in an increase of average luminous efficacy from about 45 lm/W to 54 lm/W in the decade from 1975 to 1985, [2]. The higher initial costs of electronic supplies compared to traditional ones have to be earned from a better system efficiency. For example for fluorescent lamps the electrical input power for constant light flux can be reduced by approximately 25%, partly owing to the improved luminous efficacy, partly to the reduced ballast losses. It remains, however, questionable whether primarily the resulting pay back time will determine the market penetration of the new electronic lighting systems or the additional advantages of electronically supplied lamps.

II. Potential impact of electronic supply and control on lighting systems

In addition to basic functions of the ballast, which are given in the introduction and which can be performed by traditional magnetic ballasts as well, electronic supplies are principally qualified to provide the following supplementary benefits.

* Choice of operating frequency - with restrictions, mainly imposed by the load and environmental conditions, to be discussed later. The consequences of this additional degree of freedom are manifold:
  - The light output increases for fluorescent lamps (10%, f>10 kHz) and for low pressure sodium lamps (15%, f>100 kHz) owing to a reduction of anode fall and an increase in column efficacy.
  - Operation of gas discharge lamps from DC sources.
  - Suppression of audible noise, remember the annoying 100/120 Hz noise of magnetic ballasts.
  - Reduction of ballast weight and volume owing to smaller storage elements.
  - Multiple usage of components (ignition, steady state) by means of frequency tuning. Even dimming is possible employing frequency tuning, e.g. for fluorescent lamps: 28...45 kHz, 100...25%.
* Improved system efficiency.
* Reduced weight, volume; outstanding power density of 2 kW/dm^3 possible, [5]
* Instand light without flickering.
* Operation without cathode flicker.
* Possibility of eliminating stroboscopic effects.
* A broad variety of relevant light parameters can be controlled, even remote control is applicable.

- Control of ignition and start-up phase, [6].
- Monitoring and maintenance prediction.
- Adaptive ballasts with respect to lamp type; from electrical quantities during the start-up phase, the lamp type might be recognized and properly energized.
- Light level control, even adaptive e.g. to the daylight level.

* Improved lumen maintenance of hf fluorescent lamps.
* Control of harmonic mains currents and radio interference, thus reduction of passive line filters.
* Improvement of mains power factor.
* Protection against high voltage peaks in the mains.
* Operation of electrodeless discharge lamps.

The many advantages of an hf-ballast can only be utilized effectively, if the concept of the ballast is matched to the characteristics of the lamp as well as to the requirements on the light system. The following emphasizes system architectures and working principles which have turned out to be effective.

How has the system, which provides the properties listed above, to be designed? What are the environmental conditions which have to be simultaneously taken into account?

Fig. 1 shows the system under consideration. Obviously the high frequency operation might affect the radio reception, infrared remote control systems or any electrical equipment using radio frequencies. Both the lamp itself and the ballast principally radiate radio interferences the treatment of which is presented in chap. III.

The lamp power is shaped according to the optimal operation of the individual lamp type, thus requiring different classes of converter circuits, see fig.2. The power can be transmitted continuously or pulsed. Depending on the mode of power conversion in the lamp the ballast has to provide energy storage means. This becomes immediately obvious if one remembers that the energy absorption from the mains pulsates with a frequency twice the voltage or current frequency. The voltage/ current-rating of the circuit can be reduced by decreasing the phase shift between voltage and current for given wattful power. This is the way in which the power factor improvement is performed by the electronic ballast; magnetic ballasts employ an additional capacitor in parallel to the mains for this purpose.

But the electronic ballast does not draw a purely sinusoidal current, neither does the lamp itself. The superimposed harmonics (50/60 Hz... 2 kHz) cause trouble as they are fed into the mains. The public mains is increasingly loaded with non-linear loads, thus severe regulations have been submitted. Chap. III summarizes the means to suppress or even to avoid mains harmonics.

The internal structure of the electronic ballast reflects the twofold task, see fig. 1: the input current is shaped by the "Filter" and "Preconditioner", the combination of both sometimes is referred to as "Line Conditioner", and the output power is adapted to the lamp by the "Power Output Stage", which comprises an inverter or pulseformer functionally interlinked with the ignition means.

II.1 Converter topologies and operating frequency

The choice of appropriate circuit topologies for the above mentioned functions depend on input and output voltage range, power level, operational frequency, efficiency requirements and regulations concerning electromagnetic compatibility.

All of them have to be designed and operated to meet the fundamentally necessary preconditions for the miniaturization of power converters. These are a very high switching frequency and an excellent efficiency. These properties lead to small storage inductors and capacitors, and to low heat generation, which prevents an excessive temperature rise in a low circuit volume. The temperature condition is intensified by the dissipation of the lamp. The component temperature strongly affects the operational life of the ballast.

III. Electromagnetic compatibility

The generation of both conducted and radiated emissions (Fig. 1) from ballast and lamp is a matter of major concern for all designers [9],[10]. Whereas the radiated emissions of the ballast can be eliminated or reduced to acceptable levels by screening, earthing, careful routing of cables or circuit layout, the interference problems due to the conducted emissions could only be solved in the past by additional passive filter components. As the total size and cost of the electronic supply is considerably increased by those filter components, a lot of efforts have been undertaken to optimize the fiters or to find out alternative solutions: duty cycle control by a preconditioner, frequency tuning, insertion of current-sources which compensate hf currents (active filtering); [11],[12].

With regard to radiated emissions the curves a and b of fig.3 show the maximum magnetic field strength as defined by regulations; see [10], VDE 0875/2 is a subset of [10]. To estimate the meaning of this limit to lamps, it has been recalculated into the current*loop-product of a magnetic dipole generating just the permissible field strength. The current loop is introduced as it is considered an equivalent to the real source being composed of discharge column and the lead-in wires of the lamp. Fig.4 illustrates the attenuation of the critical magnetic field by a transparent conductive layer and its effect on the allowable I*A-product. It should be noted expressly that the dipole model supports the insight into the phenomena of radiation - an adequate treatment relies heavily on the detailed calculation of the actual field distribution, which is still a complex task.

References

[1] Ch.Meyer, H.Nienhuis, Discharge Lamps, Philips Technical Library, Kluwer Technische Boeken, Doventer-Antwerpen

[2] R.Schäfer, Energy Efficient Light Sources, LS:4, pp.51

[3] H.Smulders, Electronics for Lamps, LS:4, pp.47

[4] Th.Hüsgen, K.Löhn, An ac/ac converter for ballasting gas discharge lamps, Power Conversion Congress, Paris 1984.

[5] M.Albach, Power Supply for 2 kW Halogen Lamps with Continuous Power Control and Sinusoidal Input Current, LS:5

[6] A.Wegener, Th.Hüsgen, Electronic Ballast for High Pressure Sodium Gas Discharge Lamp, LS:5

[7] G.Lütteke, H.C.Raets, High Voltage High Frequency Class-E Converter Suitable for Miniaturization, IEEE PE-1, No.4, October 1986

[8] K.Löhn, Low Loss Switching in Flyback and Halfbridge Circuits, European Conference on Power Electronics and Applications, Grenoble 1987

[9] IEC, Draft-Revision of Publication 555-2: "Harmonics", April 1988

[10] C.I.S.P.R.15, Limits and methods of measurement of radio interference characteristics of fluorescent lamps and luminaires.

[11] M.Albach, Conducted interference voltage of ac-dc converters, IEEE Power Electronics Specialist's Conference, Vancouver, 1986.

[12] M.Albach, An ac-dc converter with low mains current distortion and minimized conducted emissions, European Conference on Power Electronics, Aachen, 1989.

[13] K.Löhn, Modelling and optimization of inductive power components for high switching frequencies, Int. Macroeletronics Conference, München 1984.

[14] A.Wegener, M.Amato, A High Voltage Interface IC for Halfbridge Circuits, High Frequency Power Conversion Conference, 1989, Napels/Florida

Acknowledgements

This review is based on the expertise of colleagues from the Central Development Laboratories of Philips Lighting Division Eindhoven and from the Philips Research Laboratories. In particular the author thanks M.Albach, K.Löhn, H.-P.Stormberg, A.Wegener from the Philips Research Laboratory Aachen for their most valuable support.

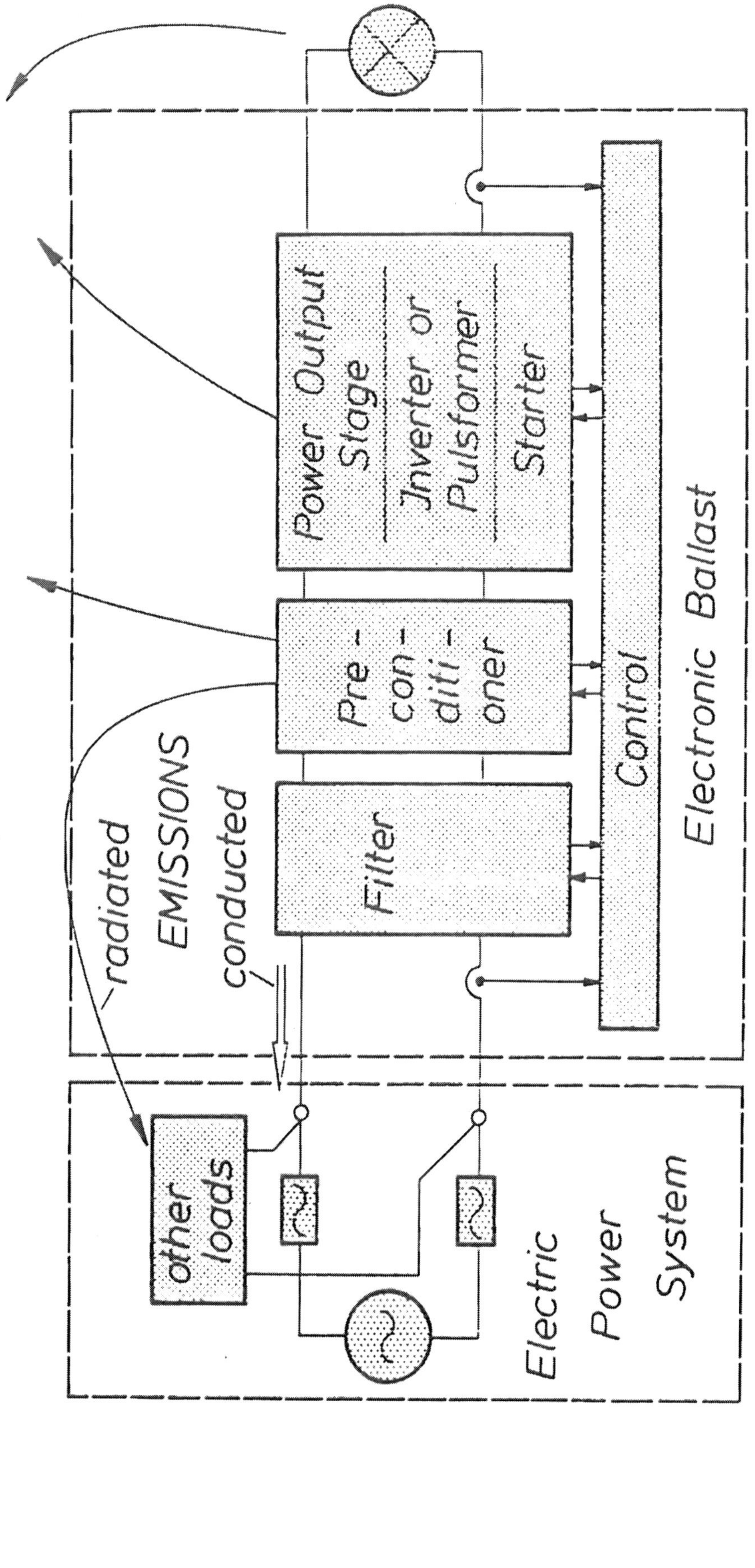

Fig. 1 Electronic Lamp System

Fig. 2 Functional Requirements on an Electronic Lamp System

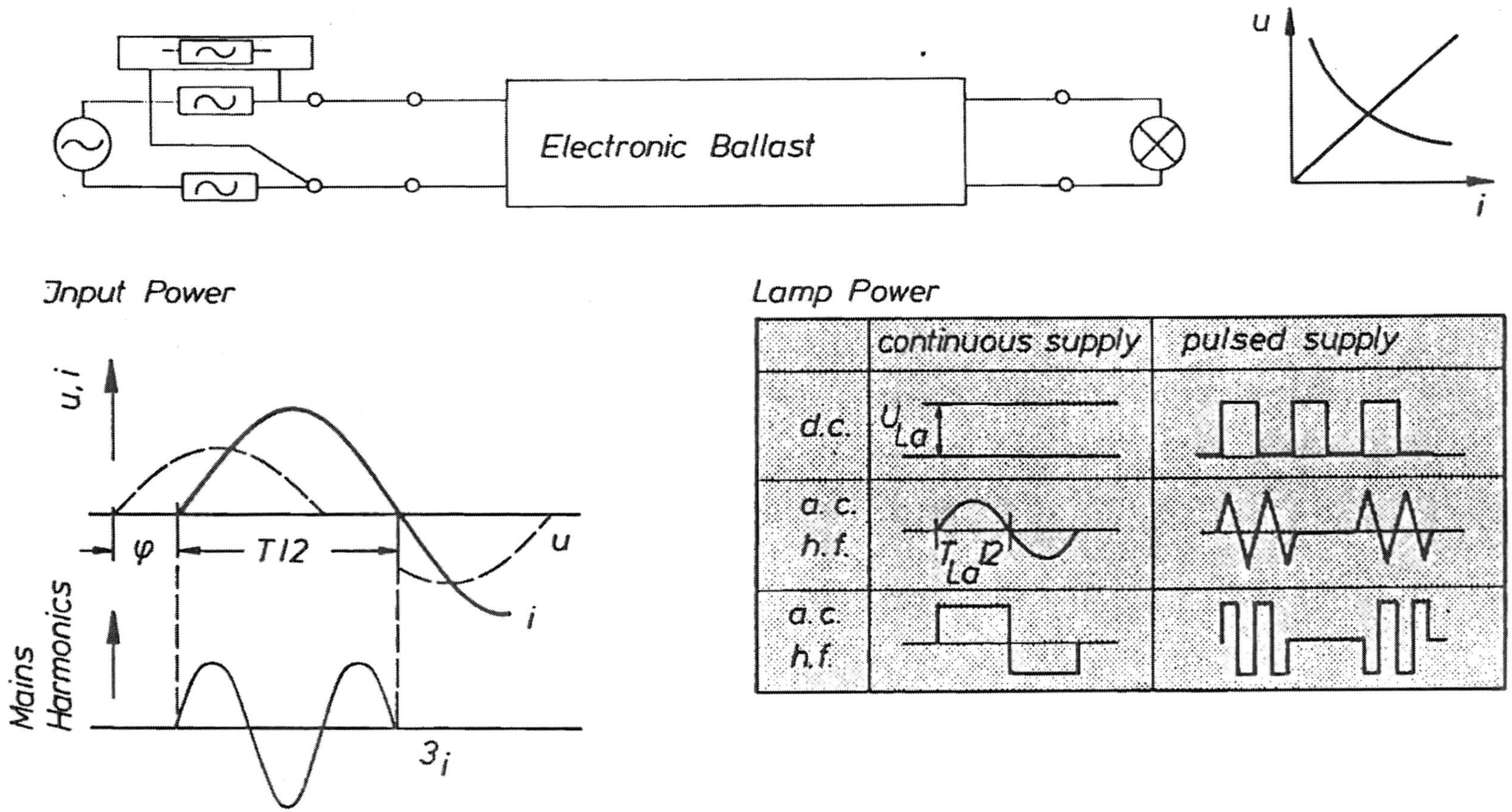

$$p_{in}(t) = UI\left\{\cos\varphi\left(1-\cos 2\cdot\frac{2\pi}{T}t\right) - \sin\varphi \sin 2\,\frac{2\pi}{T}t\right\}$$

$$P_{La} = \eta_{Ba}\cdot UI\cos\varphi$$

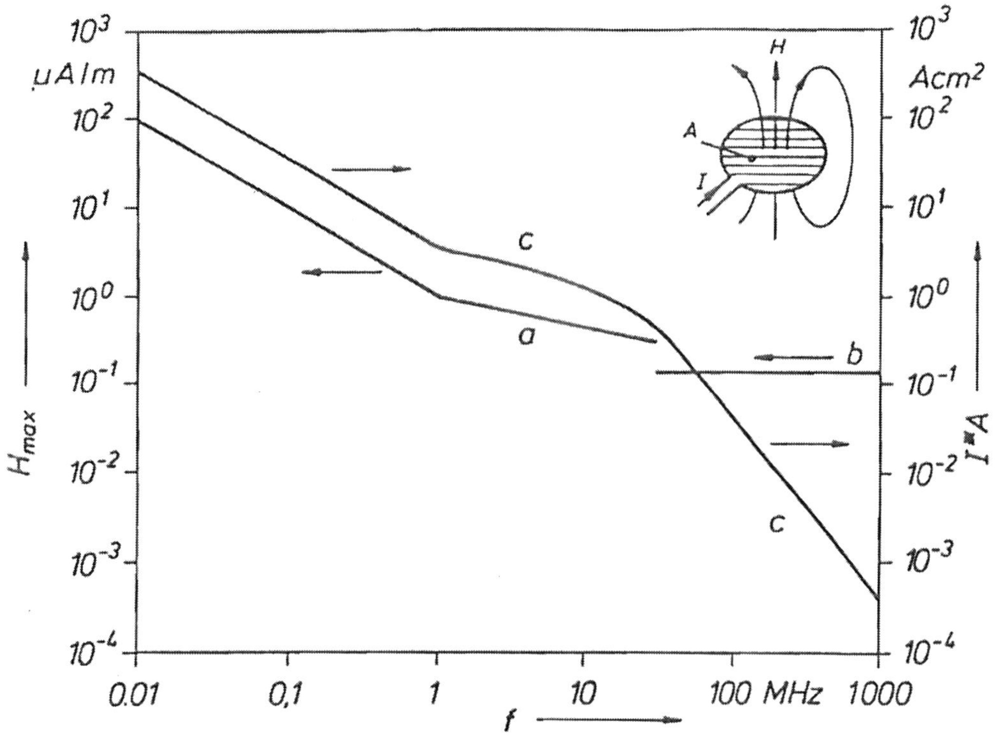

Fig.3 a. Maximum magnetic field strength, as specified by VDE 0875/2, in 3m distance.
b. Corresponding values of VDE 0871, specified for 10m distance.
c. Current*area-product of a magnetic dipole generating a magnetic field strength according to a and b, respectively.

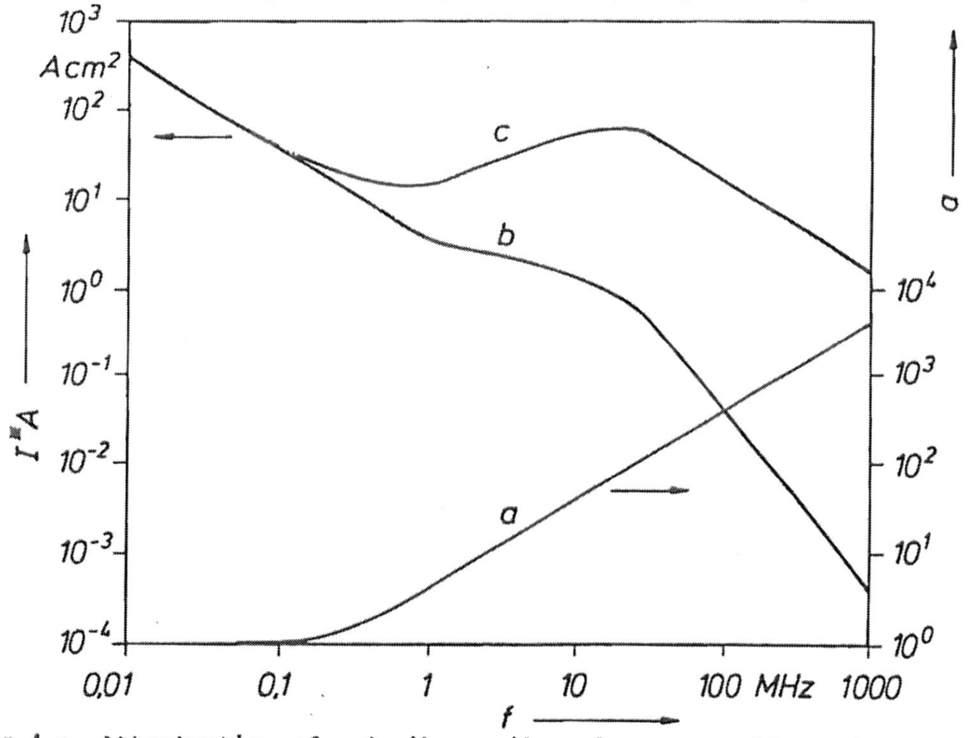

Fig.4 a. Attentuation of a 1 mikron silver layer on a 50 mm diameter lamp bulb.
b. Permissible current*area-product according to curve c of fig.3.
c. Permissible current*area-product taking into account screening of the silver layer.

An Aluminum Transport Cycle Observed by Multiphoton LIF in Operating Halogen Lamps

J. Robert Woodward
GE Lighting, Nela Park
Advanced Technology Department
Cleveland, OH 44112, U.S.A.

Abstract

The transport of Al by a halogen cycle has been observed and characterized in operating halogen lamps by Laser Induced Fluorescence. The laser system used consisted of a QuantaRay DCR2 Nd:YAg laser, scanning dye laser with R6G dye and a WEX which provided 10Hz, 10 mJ/pulse output from 285 to 277 nm. The detection system was a Spex 1 meter double spectrometer, EMI 9558QB PMT and a EG&G 4402 signal processor. The production lamps studied were 12V 50W and 12V 75W GE Precise quartz halogen lamps. The initial goal of the LIF experiment begun by Dr's. M. Kori and T. Russell was to spatially resolve the W atom concentration in an operating halogen lamp. Along with the W transition being pumped, several emission lines were detected that could not be assigned to W. These LIF signals could be eliminated by cooling the bulb wall. The resulting surface deposit was analyzed by Dr. J. Ranish with ESCA and found to contain, Si, Al, W, Fe, O and Br. This led to the identification of ten Al transitions from six upper states to the split ground state. These lines were not observed in all lamps tested and when observed their intensity varied.

Once Al lines were identified the study's goal became the characterization of the Al transport cycle. To identify the species responsible for LIF emission the spectrometer is tuned to a LIF line and various parameters systematically changed to measure the influence each change has on the LIF signal. The conditions varied were; laser frequency, laser power, lamp wall temperature, lamp filament temperature and the laser path through the lamp. The dye laser was scanned to determine the frequency absorbed by the species responsible for the Al LIF signals. Analysis of this spectrum generated the spectroscopic constants for AlBr A and X states. The Al signal intensity versus the applied laser power showed a two photon mechanism for Al LIF at filament temperatures below 2800K. The dependence of the Al signal intensity to wall temperature gave an enthalpy of 25.36 kcal/mole by the second law calculation; the enthalpy of sublimation for $AlBr_3(s)$ is 24.1 kcal/mole. The Al LIF signals were dependent on region of the lamp probed by the laser. There was a maximum near or at the wire surface which decreased to approximately zero at the Langmuir sheath diameter. The concentrations of $AlBr_3$, AlBr and Al versus temperature were calculated from available thermodynamic data using the NASA program CEC71. Comparison of Al LIF signal intensity versus filament temperature and calculated concentrations shows that the Al signal corresponds to the AlBr(g) concentration.

Multiphoton LIF of operating quartz halogen lamps has characterized the Al transport cycle as;

$$\overset{\text{Wall}}{AlBr_3(s)} \longleftrightarrow AlBr_3(g) \longleftrightarrow AlBr_2(g) \longleftrightarrow AlBr(g) \overset{\text{Filament}}{\longleftrightarrow} Al(g).$$

The two photon mechanism responsible for the Al LIF signals is;

$$AlBr(g)\ X \xrightarrow{h\nu} AlBr(g)\ A \xrightarrow{h\nu} AL^{*} \longrightarrow Al^{o} + LIF.$$

FTIR Diagnostics on Tungsten-Halogen Lamps

LAURENCE BIGIO
General Electric R&D Center
P.O. Box 8, Building K1-4C34
Schenectady, N.Y. 12301

Fourier Transform Infrared (FTIR) spectroscopy is used to measure CH_3Br, HBr, CH_4, and CO in glass halogen 90 Watt light bulbs. Measurements were made as a function of lamp burn time as well as methyl bromide concentration. Lamp burning is defined as lighting the bulb at 120 V (90 Watts) and 60 Hz. The bulbs are nominally filled with 95% Kr, ~5% N_2, and small amounts of CH_3Br ranging from 0.03-0.32% at total bulb pressures of 4000 Torr. Some bulbs studied also contain small amounts of phosphorous which coats the inside wall. The bulb inner diameter is 1 cm and its length is ~3.2 cm from base to tip-off, with the filament centered parallel to the long axis. The spectra were all taken with the bulbs unlit and at room temperature. First, an absorption spectrum is taken before a bulb is ever lit, thus serving as a reference spectrum with which to subtract out the highly wavelength-dependent glass absorption. Then, the bulb is burned for varying lengths of time after which the bulb is allowed to cool back to room temperature before another spectrum is taken. It is assumed that no chemical reactions occur during the "off" times of the bulb. Indeed, no changes in the above species concentrations were detected as a function of long (several weeks) idle sitting times.

In subtracting the pre-burn absorption spectrum from the after-burn spectra, the methyl bromide comes in as a negative going signal, whereas all the other species are positive going since they don't exist until the bulb is lit. Methyl bromide is measured by the ν_4 $^RQ_0$ band-head (at ~3060 cm^{-1}), HBr by its R(3) peak height (2621 cm^{-1}), CH_4 by its ν_3 Q-head (at ~3016 cm^{-1}), and CO by its R(7) peak height (at ~2173 cm^{-1}).

An example of the results is given in Figure 1 which shows absorbances for three of the species as a function of lamp burn time for short times. This data was taken for a bulb filled with 0.32% CH_3Br.

These signals are not yet calibrated and so only give a species population relative to itself. The error bars here are ≤6%. Evident from the figure is the short lived intermediary species, CH_4. The CH_3Br and the HBr have essentially identical decay and growth rates, respectively. The logarithms of those signals both yield first order rate constants $k=1.4\ sec^{-1}$, suggesting that very little HBr is generated by higher order reactions involving two methyl bromide molecules or one methyl bromide and one hydrogen molecule. The fact that CO is seen within 10 seconds of burn time (not shown in Figure 1) strongly suggests that the sources of oxygen are oxide contaminants in the bulb, and not from water diffusing through the glass wall.

Longer term results show a peak for CO at ~2 hours of lamp burn time, and a slow decay for HBr. Assuming a first order decay process for the HBr, one gets a rate constant $k_1 = 9.2\times10^{-4}\ hour^{-1}$. This corresponds to a half-life of 753 hours. These and other results will be presented and discussed.

Figure 1

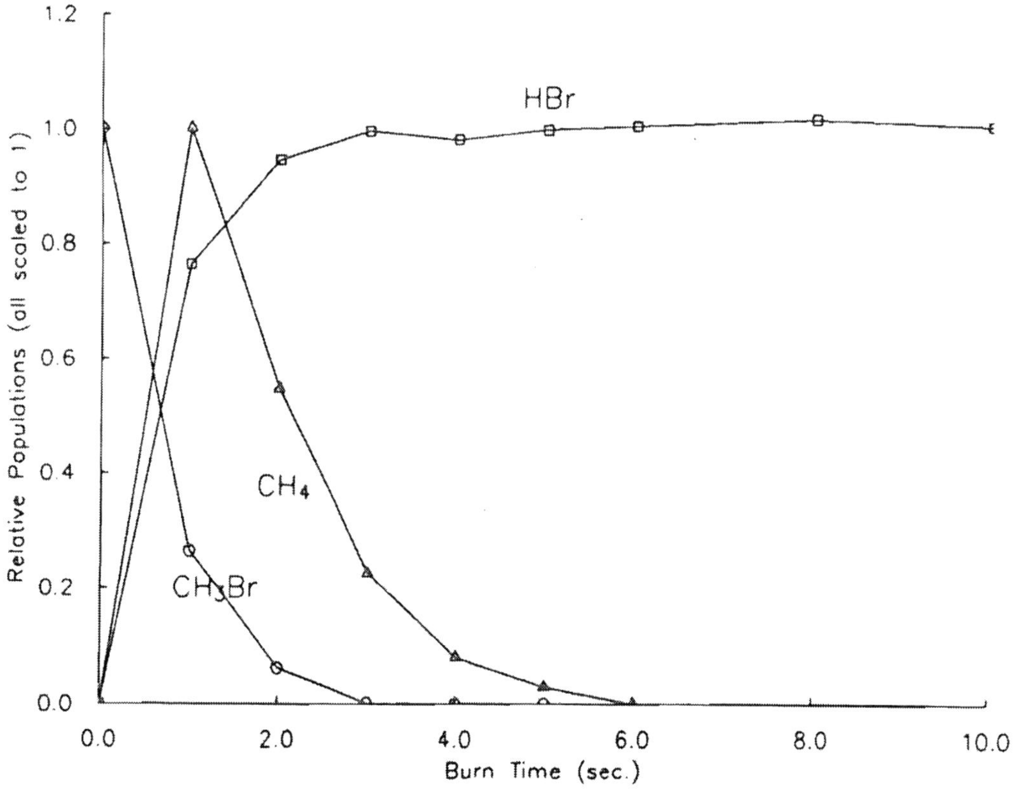

THE INVESTIGATION OF GETTER-GAS INTERACTION IN GAS FILLED INCANDESCENT LAMPS

É. Biró, J.Ádám, S.Reznik, O.Lojka[x]
TUNGSRAM RESEARCH CENTER, BUDAPEST (HUNGARY)
[x]TUNGSRAM AUSTRIA A.G., WIEN (AUSTRIA)

Getter (e.g. P or Zr) interaction in gas filled incandescent lamps with gas impurity (e.g. O_2) was studied in sealed off lamps and in modell experiments where the lamp conditions were simulated (i.e. the studied gas was Ar-N_2, with low impurity content). Measuring methods were developed based on high sensitive gas analysis.

The aim of the work was to develop rather simple measuring methods for getter qualification, which methods can be used also at the production line for the quality controll of the gettering; to determine the optimal gettering parameters (getter quality, quantity, technology) for a given lamp type; and to get some general information about the sorption-desorption properties of getters.

Measuring methods

In our laboratory mass spectrometry (MS), gaschromatography (GC) and a non destructive oxygen analyser (NDO) is used to analyse the composition and impurity content of gases, and special analytical techniques were developed, for the destructive and non destructive analysis of gaseous impurities in lamps.

The NDO e.g. is a non destructive analytical method by

means of which the oxygen content of sealed off $Ar\text{-}N_2$ filled lamps can be measured in the some ppm-some 100 ppm concentration range. This method was developed initially for the automatic selection of the leaky bulbs at the production line.

The developed sensitive gas analytical methods could be well utilised in our getter experiments. By means of the NDO technique the getter induced changes in oxygen impurity concentration could be analysed in sealed off lamps during the lamp processing and operating life continually. By means of the GC method, the getter interaction with other (different from O_2) impurity gases could be investigated in modell experiments. GC analysis was used to verifye the results of the NDO method (i.e. to check the selectivity for the O_2 determination).

Experiments and results

Incandescent lamps taken from the mass production, lamps without getters, and experimental lamps with different getters were measured. The P getter - O_2 impurity interaction was in the focus of our work. By means of the very simple NDO method a great number of O_2 analysis was made, and several getter parameters could be studied, systematically.

Some new informations about the sorption-desorption properties of getters; the measuring method used for getter qualification in the production of lamps; and data concerning the determination of the optimal getter parameters are presended. The gettering properties of different getters (red P, P_3N_5, Zr) are compared.

FLUID FLOW MODELLING AS AN AID IN THE DESIGN OF INCANDESCENT LIGHT SOURCES

D.G.Green and R.Devonshire
Chemistry Department, Sheffield University,
Sheffield S3 7HF, U.K.
S.A.Mucklejohn and N.W.O'Brien
Lamp Research, THORN Lighting Limited, Light Sources Division,
Melton Road, Leicester LE4 7PD, U.K.

Important concepts from fluid dynamics have been incorporated into the design of gas-filled incandescent light sources from the earliest stages of their development. For example, the rôle of heat dissipation by natural convection in the gas filling has been discussed in terms of characteristic numbers, i.e., the Nusselt, Grashof, and Prandtl numbers,[1] and the Langmuir sheath model[2] has proved of lasting value in predicting the heat loss from an incandescent filament to the surrounding gas.

The advent of fast, powerful computers has meant that the equations describing fluid flow (the Navier-Stokes equations) can now be solved quickly and accurately at reasonable cost. We have explored the use of fluid flow modelling in conjunction with extremely simple representations to aid the design of incandescent light sources.

This work was undertaken to assess some of the capabilities and limitations of a commercially available finite difference code for the simulation of fluid flow and heat transfer when it is applied to the modelling of commercial tungsten halogen lamps.[3] The studies were based on the THORN Lighting 500W HX121, Marine Beacon, and 750W EGG, Theatre Spotlight, tungsten halogen lamps and the calculations performed on an IBM 3083 mainframe computer using a grid size of up to 10,000 nodes.

The paper will discuss how important structural features of these lamps, namely the envelope geometry, filament geometry, filament supports and lead-throughs, and the silica bridge, can be accommodated in the necessarily simplified two- or three-dimensional models.

The study revealed the presence of complex recirculation flows in the gas filling and the marked changes in these gas flows which may be induced by changes in lamp operating conditions. The gas flows in the region of the filament were found to be particularly sensitive to operating conditions.

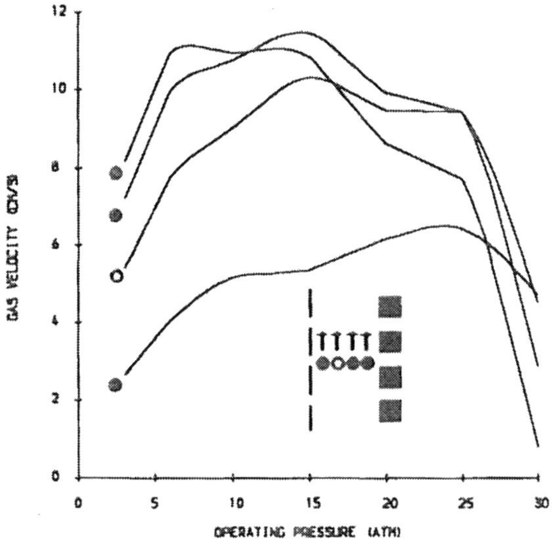

Figure 1a

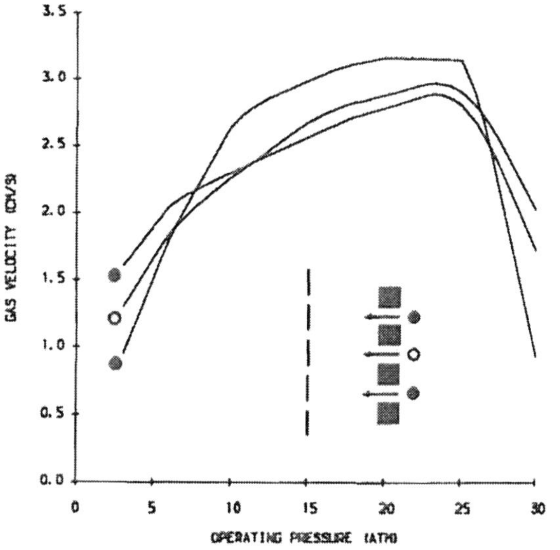

Figure 1b

Figure 1 shows the operating pressure dependence of gas velocities a) along the coil axis, and b) between the coils themselves, obtained in the study of the HX121 lamp.

References

[1] W.Elenbaas, "Light Sources," MacMillan, London 1972.

[2] I.Langmuir, Phys.Rev, 34, 40 (1912).

[3] We used the code FLUENT, Version 2.95, from Creare R&D Inc.

58:P

NON CONTACTING PRECISION MEASUREMENT OF THIN WIRES

K. Guenther, H. G. Kloss,
Kombinat NARVA Berlin, DDR
W. Luedge,
Zentrum fuer Wissenschaftlichen Geraetebau, AdW der DDR

Light source industry requires many million meters of wires every day, especially for filaments of incandescent lamps and of fluorescent lamps. Among their numerous features the geometrical shape and the diameter of the wires play an important role and must be kept within very narrow limits. Because of its accuracy and reliability the generally accepted procedure to determine the diameter is weighting a piece of 20 cm wire length. Nevertheless, this method is time wasting and destructive; it cannot be integrated into the technological process.

Optical methods seem to be suitable to solve the problem: They are non-destructive for the wire, non-contacting and allow the computerized image processing by use of CCD line detectors. However, they could not be introduced into production of wires for light sources until now since imaging systems suffer from principal physical limitations of the accuracy by diffraction, especially for thin wires. Moreover, the wire must be positioned in front of the optics with an accuracy which is unacceptable for production control.

The modern method of laser scanning derives the wire diameter from the time of interrupting the beam by the wire during the scanning cycle. The lower limits of wire diameter and error are determined by the minimum diameter of the scanning beam which is influenced by diffraction effects in the same way as in direct image processing.

The limitations described above can be cancelled by using the limiting effect itself - the diffraction - as a source of information. A wire illuminated by parallel monochromatic light produces a periodic diffraction pattern which can be focused on an optical CCD detector. It has to be noted that the significant information is given by the angle distribution of the pattern; in this way the method is not sensitive to displacements of the object. The signal is digitized and processed using FFT techniques to obtain the frequency of the diffraction pattern and the corresponding wire diameter with high accuracy.

The advantages of the diffraction method are as follows:

- contactless, simple and robust construction, no moving parts
- non-destructive measurement, suitable for process control
- both high accuracy and low requirements concerning the position of the wire within the system
- high redundancy of the information transfer between object and measuring system will result in high reliability

Technical data:

range of measurement:	10 ... 500 um	
instrument error:	> 50 um:	0.2 %;
	< 50 um:	0.1 um
measuring rhythm:	1 Hz	

required accuracy of positioning the object:

	location:	0.5 mm;
	angle:	1.5 grd

INTERACTION OF HEATED TUNGSTEN WITH ITS RADIATION

A.S.Ivantzev, M.I.Obmajikina,
V.I.Konjashkina, USSR

Non-coherent optical radiation falling on a metal results in its temperature increase. Limited resultant temperature T_{max} is calculated by known relation:

$$T_{max} = T_a + \frac{\alpha E_e}{\varkappa S}, \qquad (1)$$

where: T_a-ambient temperature;

α -absorption coefficient of metal;

E_e-metal surface irradiance;

$\varkappa$ -coefficient of thermal conductivity by convection;

S -metal surface area involved in heat exchange between metal and environment.

Interaction of heated metal with its own radiation at proper spectral interval is of interest. For instance, tungsten is heated to 2600K, environment atmosphere is argon, inner boundary surface (a bulb) is coated by heat-reflecting layer, i.e.the coating that is clear in visible region and well reflecting in infrared. For complete interaction the returned radiation with tungsten the filament should be located exactly at the bulb centre of spherical or cylindrical by the axis. These conditions are satisfied in gas-filled incandescent lamp of spherical shape.

To estimate the exceeding of tungsten filament temperatureover stationary one there used relation (1). In this case T_a, the mean temperature of bulb gas, amounts to 714K, $\mathcal{X}$ =36,1 x 10^{-5}cal/cm.sec.K. Information about coefficient of heated metal absorption, in particular, tungsten, is omitted in literature. α value for tungsten was estimated within the region of 0.8-5.0 m and temperature range of 300-2600K based on laws of non-linear interaction of non-coherent metal radiation derived from relation (2):

$$\alpha(\lambda) = \frac{4\pi e^2 \tau N}{cnm \cdot \left(1 + \frac{4\pi^2 c^2 \tau^2}{\lambda^2}\right)}, \qquad (2)$$

where: N-free electron concentration in tungsten;

e-electron charge;

τ-relaxation time of atoms;

c- rate of light;

n-index of deflection;

λ-radiation wavelength;

m-electron mass.

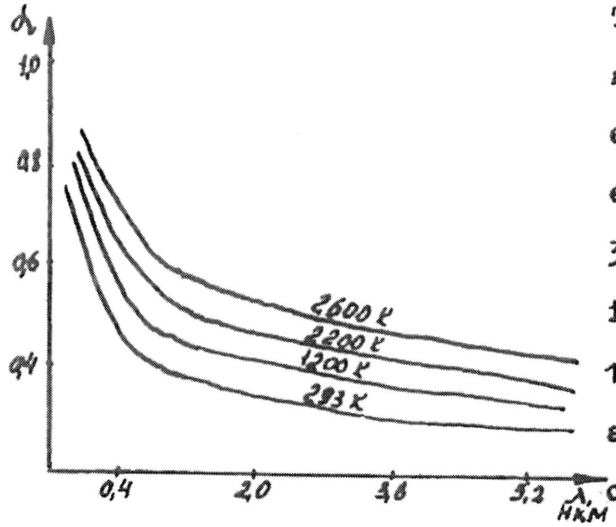

These dependences are presented in the figure. Values by expression (1) showed that tungsten wire 390 m by diameter wound in coiled-coil cylinder, 12mm long and 4,7mm by diameter, heated by electrical current to 2600K has T=2880K, i.e. additional heating by infrared radiation contributes to temperature increase by 280K. Experimental measurement gave temperature growth within 180-200K.

Temperature influence upon optical parameters of the infrared reflecting layers

T. Kanabus, W. Stanioch

Research Development Center POLAM

Karolkowa 32/44, 01-207 Warsaw, Poland

IR reflecting layers to be used in light sources for the light efficiency increasing should be resistant among others against high temperature of the bulb. Some experiments on temperature influence upon the light transmittance and IR reflectance of $InO_2/Ag/InO_2$ layers have been carried out using the cathode sputtering method and spectrophotometric measurements of the layers as well as special mathematic model of the results calculations. Three samples of layers were performed by means of routine cathode sputtering and a spectral direct transmittance in full visible range as well as the spectral direct reflectance in infrared range up to 2,5μ of optical radiation have been measured by spectrophotometer SPECORD M 40 and own arrangement.
As a criterion of the layers usefulness for lamp application the special analytic model has been used for relative light efficiency of radiation RLE(R) calculation:

$$RLE(R)=LE(R)_l : LE(R)_r$$

in which: $LE(R)_l$ – light efficiency of radiation with the layer use

$LE(R)_r$ – light efficiency of radiation without any layer (reference)

$$LE(R)_l = \frac{683.\Sigma\{M_e(\lambda).\tau(\lambda).V(\lambda)\}}{\Sigma\{M_e(\lambda)[1-/\rho(\lambda).\varepsilon(\lambda).k/:/1-\rho(\lambda)+\rho(\lambda).\varepsilon(\lambda).k/]\}}$$

in which: $M_e(\lambda)$ – spectral radiance of tungsten light source

$\tau(\lambda)$ – spectral transmittance of the layer

$V(\lambda)$ – spectral luminous efficiency CIE

$\rho(\lambda)$ – spectral reflectance of the layer

$\varepsilon(\lambda)$ – spectral emissivity of the tungsten

k – utilization factor of IR radiation reflecting (equal to 1 in the calculation presented here)

$$LE(R)_r = \{683.\Sigma[M_e(\lambda).V(\lambda)]\}:\{\Sigma[M_e(\lambda)]\}$$

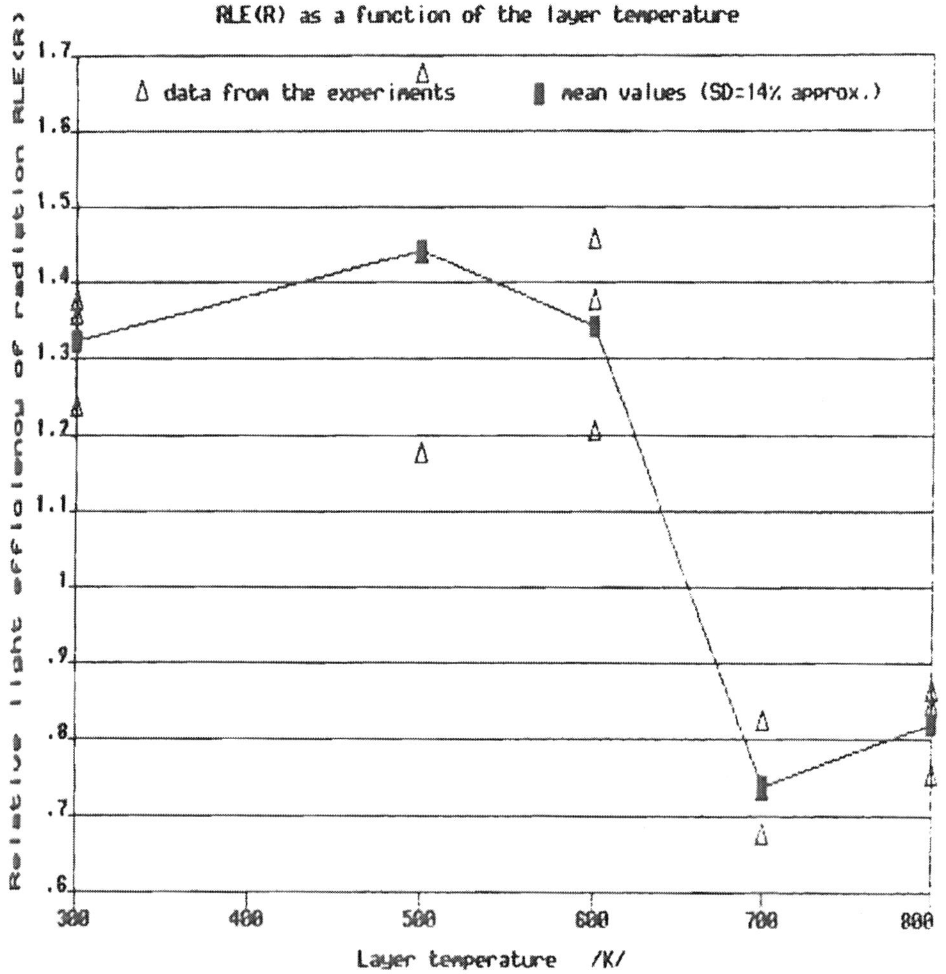

The result of the experiment and calculations made for 2500 K of tungsten distribution temperature is shown on the diagram representing the RLE(R) values as a function of the layer temperature. Each layer sample was warmed in air for 7 hours in the temperature 300 K, 500 K, 600 K, 700 K and 800 K successively and then measured by spectrophotometer SPECORD M 40 (in visible range) and by own arrangement (in IR range).
It is clear from the diagram, that RLE(R) values for the layer temperatures above 600 K shortly decrease what is resulted by IR reflectance decreasing. A conclusion is that the layer $InO_2/Ag/InO_2$ in application to tungsten light source can be used if the layer (bulb) temperature is not greater than 600 K. The same conclusion can be state if the RLE(R) is calculated for 2800 K of the tungsten distribution temperature but the RLE(R) values are in this case slightly lower.

Electric Bulb Based Radiation IR Microfurnace for Fast Sintering and Epitaxy

Andrzej Szymański *)

Maltech Int. Ltd. Co., Warsaw, Poland

The epitaxy process serves for the deposition of a condensed solid phase on a properly chosen substrate. The method called in short CVD can use for the solid layer deposition the chemical reaction takeing place between the gaseous reagents. The driving force of the process is the excess of free enthalpy, necessary for overcoming a number of kinetic barriers, such as the diffusion of reagents to the substrate, chemical reactions and other surface processes. The CVD process requires the presence of a chemical reaction for the deposition of the desired product on the intersurface. An excess of free enthalpy of the deposited material and kind of the substrate determine if the nucleation and growth of a given phase are forbidden, or if an amorphous, poly- or monocrystalline layers grow.

The diamond epitaxy process on seed is a homogeneous one, and the kinetics of chemical reactions related to the formation of a new phase, depends on both the possibilities of reaction between gaseous molecules and the character of the substrate surface. Such a process was studied. An optical ellipoidal furnace was constructed and carried out from steel, and its internal surface was covered by a mirror-like gold layer. The heat energy source was a 55 W bulb set in one of foci. In the other focus there was a table made of alumina ceramics, on which seed crystal for epitaxy were set. The device ensured the working vacuum of the order of 10 Pa. Propane-butane and propane-butane-hydrogene mixtures were applied as the working gases. The process temperature was 1120 K.

The feasibility study of the implementation of a series production of optical radiation micro-furnaces of this type showed that the optimum solution is their construction in the form of an ellipsoidal bulb lamp equipped by setting a 55 W heat energy source at the upper focus, the supply and desupply of the gas medium, the vacuum lamp terminals and the alundum ceramics worktable introduced from below. The internal surface of the bulb should be covered with a chemically resistant metallic mirror used in many tapes of electronic tubes.

---

*) 6/14 Krucza Str., Apt. 113, 00-537 Warsaw, Poland

PRODUCTION TECHNOLOGY, SOME ELECTRICAL AND LIGHT PARAMETERS OF INCANDESCENT LAMPS WITH FILAMENT OF COMPOSITE MATERIAL ON CARBON BASE

N.S.Vdovin, M.I.Rogajilin, A.V.Mordjuk, USSR

There presented work results on new material application (carbon-graphite fibres) for incandescent lamp filament body. Instead of high critical tungsten, as a filament for light sources, can be used spirals of composite materials with a carbon-graphite fibre as a filler, and crystalline pyrocarbon- as a matrix. There described some properties of initial carbonic fibres of various grades and investigation results of property changes in composite carbon-carbon materials on base of these fibres in terms of duration of pyrocarbon deposition process. It was shown that the higher pyrocarbon content in composite, the better is material strengthening characteristics and the lower - its electrical resistance.

There supposed the following technological chart of filament forming from carbon fibres for incandescent lamps:

-forming of initial spiral-shaped blanks from carbon fibre by dry wiring on quartz or graphite mandrel of definte size;

-thermochemical treatment of initial spiral-shaped blank to get composite carbon-carbon material by pyrocarbon deposition from gas phase on surface and into a blank volume by thermal deposition of carbon contained gas;

-high-heat treatment of produced filament bodies at

their operating temperature in order to relieve inherent stresses between fibre and pyrocarbon occuring at the volumetric packing process of initial blank by pyrocarbon.

There supposed a principal estimation model of technological process parameters in packing of initial filament body blanks by pyrocarbon. Discussed were some estimated and experimental data of working characteristics of carbon filament; radiation energies in visible and infrared regions, estimated by known Planck relation taking into account graphite spectral radiant emissivity, evaporation rate and high-temperature creep characteristics as compared with tungsten filament body.

Investigations on sampling lamps with carbon filament showed that at colour temperature of 2100-2200K lamp light output is 3,5-4lm/W, service life - 800-1000hours.

Lamp light output essentially increased at filament body colour temperature of 2700K but its service life became well below due to strong blackening because of large quantity of carbon atoms intensively evaporated from filament and deposited on its walls. In view of that there shown some ways for realization and proceeding the regenerative cycle in carbon filament lamps and supposed some new solutions for use of regenerative cycle in these lamps.

There given perspectives of further development of incandescent lamps with carbon filament body.

Molybdenum Blackening in Glass Halogen Lamps

Thomas H. Yu and Joseph M. Ranish
G E Lighting
NELA Park, Cleveland, OH 44112

In single ended low wattage (<60W) incandescent glass halogen filament tubes containing a methyl bromide, nitrogen, xenon fill, a phosphorus getter, and mounted within a thick glass bottle, corrosion of the molybdenum lead wire at the clamp and formation of a black deposit on the bulb wall is sometimes observed when the lamp is burnt in the base up position (with the coil body axis parallel to the gravitional force of earth). The deposits, if formed, are easily seen after 20 hours of burning. No such blackening has been observed, however, when lamps are burnt in the horizontal position. Analysis of the black deposit indicates that it is primarily metallic molybdenum with lesser amounts of bromine, oxidized tungsten, and oxidized and elemental phosphorus. Typically, the atom ratio of Mo to W is 12 to 1. Lamps similar to the above but not containing the methyl bromide slowly form a diffuse deposit containing essentially only tungsten when burnt base up. This implies that the transport agent is a bromine containing molybdenum compound. Further investigation was focused on the major differences between horizontal and base up burning: wall temperature, lead temperature, and distance between the coil body and the lead wire.

To investigate the effect of filament tube wall temperature on the transport of molybdenum, filament tubes were burnt base up, in the open air or in a furnace so that the wall temperature was maintained constant within the range, 500 K to 980 K. Observations after burning 20 hours reveal slight diffuse blackening near the pinch and a light brown film covering most of the interior of the filament tube for 500 K wall temperatures, severe blackening of the middle half of the filament tube for 600 K to 770 K wall temperatures, and no blackening for wall temperatures greater than 800 K. The diffuse blackening contained metallic tungsten and molybdenum in approximately equal atom amounts as well as phosphorus and nitrogen. The light brown film contained oxidized tungsten and oxidized molybdenum with a Mo to W atom ratio of about two as well as phosphorus, nitrogen, and about half of the bromine initially dosed into the filament tube. The severe blackening had a composition which was similar to that seen originally. The greatly reduced blackening of the filament tube burnt at a 500 K wall temperature is probably due to condensation of the molybdenum transport species on the wall and an insufficiently high wall temperature for rapid hydrogen reduction of that species.

The effects of the distance from the coil body to the lead wire and the lead wire temperature were investigated by constructing filament tubes having 1, 3, or 5 mm separations between the coil body and the clamp of the short lead wire and having short lead wire diameters in the range, 0.2 to 0.8 mm. Since the lamps were burnt base up, only the geometry in the vicinity of the topmost lead wire, the short clamp area, was altered. The lead wire diameters were chosen so that a common range of temperatures at the short clamp would exist for each of the coil body-clamp separations. One set of these filament tubes was burnt 20 hours, base up, in open air; another was burnt 20 hours, base up, jacketed within glass bottles. For the set burnt in open air, a diffuse metallic blackening near the pinch region was observed for practically all of the filament tubes with the 1 mm coil body-clamp separation, for a fraction of the filament tubes with the 3 mm coil body-clamp separation, and for practically none of the filament tubes with the 5 mm coil body-clamp separation. A light brown film covering the interior of the filament tube was also observed. The amount of this film decreased as the coil body-clamp separation increased. These observations were independent of the lead wire diameter. For the set of jacketed filament tubes, severe blackening of the middle half of the filament tube was observed for about 40% of the filament tubes with the 1 mm coil body-clamp separation, for about 13% of the filament tubes with the 3 mm coil body-clamp separation, and for about 7% of the filament tubes with the 5 mm coil body-clamp separation. Current data is insufficient to determine how clamp temperature is correlated with blackening incidence.

Attempts to understand this phenomena in terms of chemical thermodynamic equilibrium are not successful. This is not surprising given the failure of this approach to explain the observed tungsten transport in the same system. For this hydrogen rich system, neither molybdenum bromides nor molybdenum oxybromides are thermodynamically stable at any of the wall or clamp temperatures studied. The dependence of the blackening incidence on the coil body-clamp separation strongly suggests that the molybdenum clamp corrosion is kinetically controlled. Blackening then results as the flow from the clamp area approaches chemical equilibrium. When blackening is observed, the fact that not all of the lamps blacken indicates that another, as yet undiscovered, factor is involved as well. This hypothesis, currently undergoing additional scrutiny, will be discussed at the meeting.

64:P

Studies about halogen regenerate cycle agent in halogen lamp

Zhou Weizhong and Cao Shanchun
Institute of Electric Light Sources, Fudan University
Shanghai, China

Abstract

This paper relates the research results of bromine, chlorine and fluorine systems as halogen regenerate cycle agent. During operating, halogen lamp will be blacken, in order to overcomes the blackening in inner wall of the bulb and increase lifetime of halogen lamp. We choose many halide compounds as cycle agent. It go on tests in the bulb. We have obtained many useful effects.

In copying lamp, we elect $CHCl_3$ - CH_2Cl_2 or $CHCl_3$ - CH_2Br_2 systems to solve this bulb's blackening on filament support region. We obtain a optimum H/X rate and optimum wide of mixture filling content, namely H/X rate is 0.85 and $CHCl_3$ filling content is 0.3-0.7μg/ml, CH_2Cl_2 filling content is 1.3-3.0μg/ml, CH_2Br_2 filling content is 2.0-6.0μg/ml. Lifetime of copying lamp have obtained above 200 thousand number of times. The test condition of copying lamp (330W/150V) is operating time 4 sec, stop 4 sec, lamp is level position.

In order to solve the problems of avoiding the fluorine attack on the colder region of the filament and the inner wall of the bulb at higher temperatune.

We have adopted two methods.. First, find out a fluorine compound which can be used to carry out the reaction as fluorine, and at the same time must be completely inert and easily handled under normal conditions. It also must be dissociated into fluorine at higher temperature.In order to decrease the corrosion, the fluorine concentration in bulb must to reduced. If may be achieved by means of adding another additive, such as the bromine compounds.

We have used the NF_3- HBr system, second, the inner wall of bulb is coated with a layer of complex titanium aluminium phosphate. The test lamp as operated at 400W/30V (30lm/W, 3200K), The bulb is filled with a rare gas, such as Xenon (5000torr). It only is added in the bulb, The average lifetime of the lamp is 100h. If only NF_3 is filled in the bulb, the lifetime of the lamp is short and the bulb is blackened.

We select NF_3- HBr systems as the tungsten fluorine cycle agent and the inner wall of the bulb have a coated layer. The evident increase of the lifetime of the lamp is obtained 300-500h. But white spots will appear on the inner wall of the bulb. Because the coating is spoiled by fluorine.

COMPACT QUARTZ HALOGEN LAMPS WITH INFRARED REFLECTING COATINGS

R. S. Bergman
Advanced Technology Department
GE Lighting, Cleveland, Ohio, 44112, USA

It is well known that the efficacy of incandescent lamps could be greatly increased if the radiated energy in the near infrared (IR) region could be returned to the tungsten filament. As a consequence lamp manufacturers have searched for a long time for coatings that would allow IR lamps to be made. To accomplish this for halogen lamps requires simultaneous development of:

- films on the lamp envelope with high infrared reflectance and excellent visible transmittance;
- compatibility of the film with the envelope material, shape and operating temperature; and
- functional optical design of the filament/envelope combination to enhance the capture probability of the reflected IR.

Lamp products using IR reflecting films have been marketed with different degrees of success dependent on which type of transparent heat reflector used.[1] For example Durotest marketed an incandescent lamp incorporating a TiO_2-Ag-TiO_2 thin film on a specially made elliptical lamp envelope.[2] While this film gives excellent infrared reflection it is not suitable for halogen lamps high temperature conditions and the visible transmission, i.e., color characteristic, is much less than desired.

In 1981 GE introduced a line of linear quartz halogen lamps using a multi-layer dielectric film to achieve efficacy increases of 30 to 40%.[3] This film is compatible with high temperature materials and provides good color. However the film application was restricted to long cylinders. With the introduction of a 60 Watt halogen/IR PAR 38 lamp at the 1989 Lighting World a compact, low wattage, line voltage halogen source using transparent heat reflectors has been achieved.

The choice in design, limited by constraints of effectively using IR, especially as related to "display" applications, will be considered in detail. The role of modeling in the initial design stage is accented. Major elements of the design chosen, with achieved product performance, is summarized as follows. The IR film is multi-layer refractory oxide whose design attempts to optimize color uniformity. Measurements show achievement of color uniformity within a two-step oval, centered near the blackbody locus, with a shift of about -30 to 50K in color temperature with respect to non-coated sources.

A principle motivation in arriving at the structure chosen was the achievement of rather precise radially centering of the filament within the IR coated envelope over life. This necessitated a double-ended self-centering mount construction and design of a sag resistant compact coil-coil filament coupled with an elliptically shaped quartz envelope. The resulting product provides an efficacy increase of over 30% compared to similar non-IR products with variations in power, lumens and life nearly as good as present halogen lamps. In addition the compact coil provides advantages in beam applications. Finally, the advantages/disadvantages of using quartz as the envelope material will be considered.

References

1. H. Kostlin & G. Frank, " Thin-film reflectors", Philips Tech. Rev., 41, 225-38 (1983/4)

2. I.S. Goldstein, R.P. Fontana, L. Thorington & R.P. Howson, "The Design, Construction and Performance of an Incandescent Light Source with a Transparent Mirror", Light. Tech., 18, 93-7 (1986)

3. L.E. Hoegler & T.K. McGowan, "Practical High Efficiency Tungsten Halogen Lamps using IR Reflecting Coatings", JIES, 165-174 (1984)

REVIEW OF OPTICAL COATINGS FOR INCANDECENT AND OTHER LAMPS

Y. YUGE
TOSHIBA LIGHTING & TECHNOLOGY CORPORATION
1-201-1 FUNAKOSHI YOKOSUKA 237 JAPAN

Introduction

It can be said that the trend of the development of light sources up to the present day has been the development toward the improvement of luminous efficacy.

A major increase of efficacy owes a great deal to new materials such as tungsten filaments of incandescent lamp, phosphors of fluorescent lamp, metal halides of high intensity discharge lamp. The material for optical thin film and its application technique would also be important from the view point of the luminous efficacy improvement.

The applications of optical thin films to light source can be classified into the following three categories.

(1) Thin films for diffusing light.

(2) Thin films for improving the luminous efficacy of light source.

(3) Thin films to reduce unnecessary radiant energy.

The techniques of improving the luminous efficacy of light source by applying optical thin films transmitting visible light while reflecting infrared rays are most interesting, and their application to incandescent lamps has long been investigated.

For low pressure sodium lamp, the infrared reflective film was first applied in 1960s, and the luminous efficacy was increased to a large extent.

In this paper, recent trend in development and coating technologies are discussed with respect to the optical material focussing mainly on the infrared reflective material for the incandescent lamp.

To reduce the radiation losses in infrared, the coating to be applied must therefore be effectively reflect the infrared of wavelength greater than about 2.5 μm.

It is already known that substances containing a fairly high concentration of free charge carriers and having a large mobility of charge carriers will be good infrared reflectors. Semiconductor materials, SnO_2 has long been used as transparent electrodes and having infrared reflective function, therefore came to be applied for IR reflective films of low pressure sodium lamps.

In general, the SnO_2 film is formed by spraying such material as $SnCl_4$ or by CVD method.

Moreover, one of the sorts of material, capable of increasing free charge carrier concentration and mobility, is In_2O_3, which is transformed into n type semiconductor by reduction and/or by doping of impurities such as SnO_2, F.

For the coating methods of films, Pyrosol process and the like other than spraying method are employed.

Fig. 1 shows the typical spectral transmittance and reflectance of In_2O_3(Sn) film and SnO_2(F) film. The optimizations of In_2O_3 films are still under way, and a high luminous efficacy of 200 lm/W with 130W low pressure sodium lamp has been reported. (10...16)

(b) Incandescent lamp

Luminous flux can easily be controlled by PAR (Spotlight) type lamps in which the parabolic reflector and the lens are fused together. This type of bulb is currently used in storelighting, floodlighting, and its utilizability is high.

Its aluminum deposited reflecting surface however reflects also infrared rays. And so in order to reduce quantity of infrared rays radiated forward, a bulb having front lens coated with infrared reflecting materials such as In_2O_3(Sn), SnO_2(Sb), etc., has been realized. Other type of bulb, having In_2O_3(Sn) coating on the inner surface of its front lens and over-coated with another antireflection film of SiO_2, can reduce transmission of infrared rays to 30 % or lower with a slight decrease in the luminous output. (8)

(3) Application of multilayer films

(a) Cold mirror

This type of films utilizing multiple optical interferences realized since long ago consists of alternately formed layers of ZnS films (dielectrics having high refractive index) and MgF_2 films (dielectrics having low refractive index). Almost all of this type of films are formed by vacuum deposition.

Utilization of multilayer films in the form of a so-called cold mirror, replacing aluminum deposited films, effectively reflecting visible light and transmitting infrared rays has remarkably reduced the forward radiation of infrared rays. A lamp in which tungsten halogen lamp is integrated with a cold mirror is utilized to a great extent at present not only in storelighting but also for various projector lamps. It radiates 80 % or more of infrared energy backward and hence, the heat radiated forward is minimized.

For films consisting of alternately formed layers of ZnS and MgF_2 by vacuum deposition, various effort has been made to improve its heat resistance. For developing these films having higher heat resistance, other materials (TiO_2 - SiO_2) and techniques have been extensively investigated. (17)

(b) Incandescent lamp and Tungsten halogen lamps

With respect to the application of IR reflective film to lamps, it has been reported that 3 layer films TiO_2-Ag-TiO_2 could improve the efficacy or conserve 60 % of electric power of incandescent lamp. ZnS-Ag-ZnS films and coating technique of vacuum deposition or sputtering are also studied. In the experiment, by optimizing lamp structure, filament design, an electric energy conservation factor approximately equal to the above-mentioned value(or 60 %) has been obtained. But due to many problems related to production techniques, the industrialization has not yet been realized.

Figs. 2 shows the characteristics of IR reflective films.
(3...6,8,9,18...21)

These days, new types of lamps are realized in which IR reflective films are formed on the outer surface of the sili glass bulb of tungsten halogen lamp to increase the lumino efficacy. IR reflective films having heat resistance of 500°C o higher are required in a tungsten halogen lamp. For th purpose, multilayer films of TiO_2 -SiO_2 or Ta_2O_5 -SiO_2 ar practically used. In the case of the multilayer films of Ti and SiO_2 realized in Japan, the thin films are formed by dippi technique using organometallic solutions instead of th conventional vacuum deposition, sputtering, etc. The method which glass bulb is dipped in organometallic compound solutio such as metal alkoxide and the like, on which detail researches were made by H. Schroeder in 1969. Other than th films of TiO_2 and SiO_2, various films of such material as SnO In_2O_3 are obtained by this method, and their applications cold mirrors and laser mirrors have been proposed. However, was only in recent years that they were applied to tungst halogen lamps. The success of this application owes much on th results of the improvements in the alkoxide solution enablin formation of optical multilayer films and also in the methods o coating.

Fig.3 shows the formation of thin films from organometalli solutions. A glass bulb is dipped in the coating solutio containing metal alkoxide, then pulled up at a specifi constant speed to form gel-state films containing metal alkoxi on the surface of the bulb and baked at a high temperature t form rigid films of metal oxide. Although the complex reactio of metal alkoxide in the pulling up process is not fully clear, it is thought to be due to the hydrolysis reaction (1) with H_2 contained in the environment during pulling up process and al polymerizing reaction.

In the films after the completion of pulling up proce hydroxyl groups and alcohol(ROH) are formed. Next, the gel-stat films are baked at a high temperature to form rigid metal oxid films due to the vaporization of alcohol and the dehydrating condensation of hydroxyl groups. (2)

$$M(OR)n + nH_2O \;---\; M(OH)n + nROH \qquad (1)$$

$$M(OH)n \;---\; MOn/2 + n/2\,H_2O \qquad (2)$$

where M is a metal and R is a alkyl group.

The thickness of formed thin films is influenced by such factors as viscosity and concentration of solution, the speed of pulling up, the environment during the pulling up process, and the baking conditions.

Fig.4 shows typical spectral transmittance of TiO_2-SiO_2 films (with 16 layers) coated on a quartz tube.

In the lamps currently realized, conservation of electric power of 15 % and 20 % have been accomplished for single ended and double ended tungsten halogen lamps, respectively. By combining these lamps with cold mirror (dichroic mirror), it is made possible to reduce the forward heat radiation by approx. 90 %, and these lamps are widely utilized in the storelighting in Japan. (2,7,22,23)

Glass plate has been the generally used substrate for the cold mirror, but similar effects can be obtained by using metallic aluminum substrate on which multilayer film of TiO_2 and SiO_2 is formed. A cold mirror made by metallic substrate is realized for tungsten halogen lamp.

In the case of high wattage tungsten halogen lamp, because of the requirement for higher heat resistance, multilayer films of Ta_2O_5-SiO_2 are used. It is realized by vacuum deposition method, however, for homogeneous coating of films in the longitudinal and radial directions of a glass tube, means such as rotary mechanism and the like are employed in the process.

Lumen output of 1500W lamp is obtained by a lamp of 900 W. This corresponds to 40 % energy conservation.

Since the geometrical reflection factor (which means the returning rate of infrared rays radiated to the tungsten filaments) is influenced by the shapes of bulb, filament, etc, comparison of film characteristics of different types of lamps cannot be made only by the rate of the conservation of electric energy. The IR reflective films practically used in lamps still have room for improvement with respect to their optical performances or costs, and extensive efforts are being made. (24,25,26)

(c) <u>Other lamps</u>

Another example of IR reflective films application is the improvement of high pressure sodium lamps used for plant growth. The PAR (DIN) (photosynthetically active radiation) is increased by returning the infrared rays back to the arc tube by TiO_2-SiO_2 films coated on the outer jacket, and hence the heat radiation is reduced.

Fig. 5 shows the structure of the test lamp. (27)

Future

As is described in the beginning of the present paper, the development of light source owes its success to that of material to a great deal. IR reflective films have begun to be applied to lamps and tungsten halogen lamps with IR reflective films have now commercially available.

Further, efforts will be made to improve heat resistance, the optical characteristics, and to lower the cost. The IR reflective films will be applied not only to incandescent lamps and tungsten halogen lamps, but also to other light sources.

It is considered that the techniques of optical thin films controlling the radiated rays would be one of the most important keys for the future development of light sources.

<u>REFERENCES</u>

1. I E C 34A(Secretariat) 320 , May 1988
2. A. Kawakatsu, Y. Yuge , Proc. of 1st LUX PACIFICA(Shanghai) (1989) 226
3. J.C.C. Fan et al , Proc. of 9th Intersociety Energy Conversion Engineering Conf. San Francisco (1974) 341
4. J. Brett et al , J. Illum. Engng. Soc. 9 - 4 (1980) 197
5. J. Brett et al , J. Illum. Engng. Soc. 10 - 4 (1981) 214
6. H. Kostlin , 3rd International Symposium on the Science and Technology of light sources 1983
7. N. Ichinose et al , ibid. 1983
8. H. Kostlin, G. Frank , Philips Tech. Rev. 41 - 7/8 (1983/84) 225
9. H. Kostlin , J. Illum. Engng. Soc. 14 - 1 (1984) 151

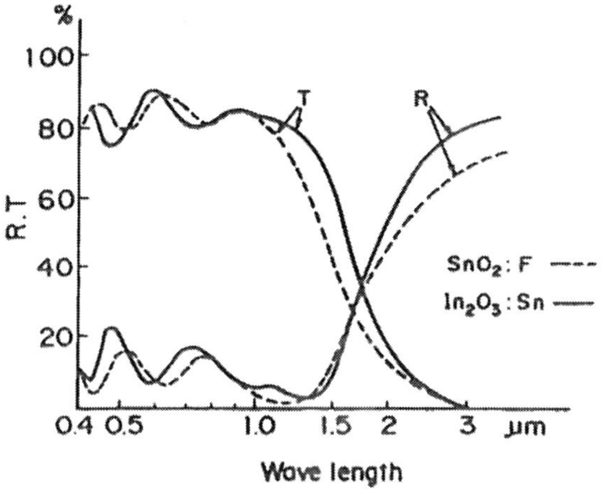

Fig.1 Typical transmittance T and reflectance R of In_2O_3(Sn) film and SnO_2(F) film on glass

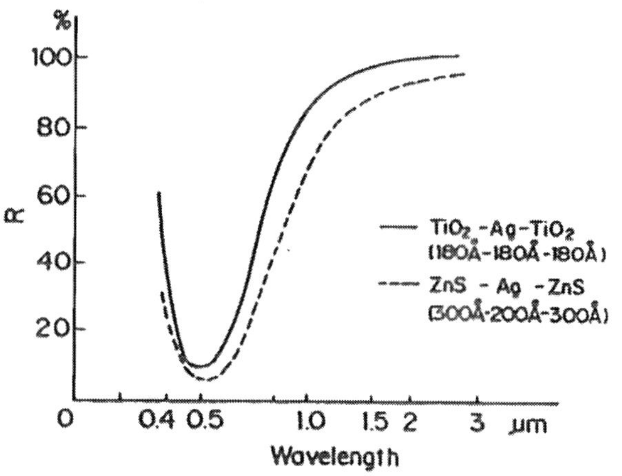

Fig.2 Reflectance R of IR reflective films consisting of metal-dielectric film

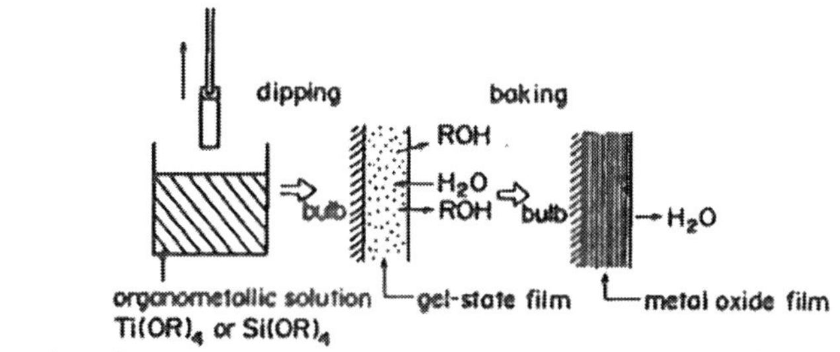

$$RO\left[\begin{matrix}OR\\ M-O\\ OR\end{matrix}\right]_n R+(2n+2)H_2O \longrightarrow HO\left[\begin{matrix}OH\\ M-O\\ OH\end{matrix}\right]_n H+(2n+2)HOR \longrightarrow \left[\begin{matrix}O\\ O-M-O\\ O\end{matrix}\right]_n$$

Fig.3 Formation of thin films from organometallic solusions

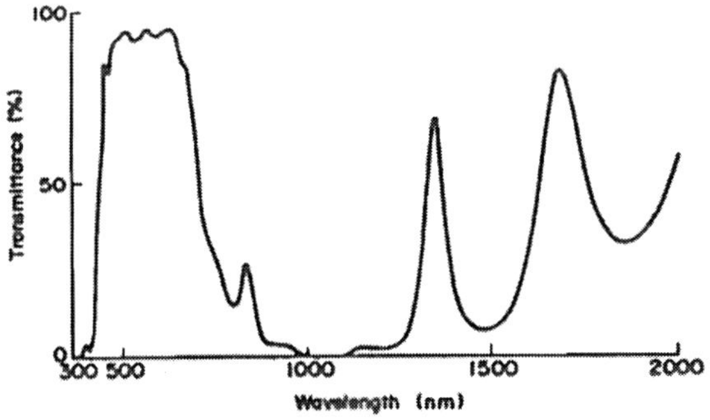

Fig.4 Typical spectral transmittance of TiO_2-SiO_2 films (with 16 layers)

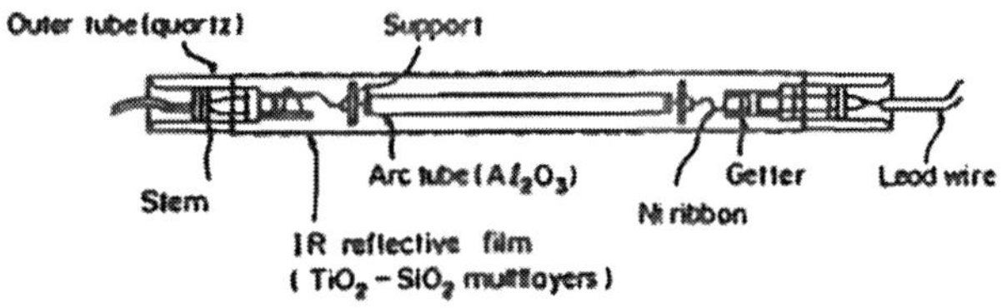

Fig.5 High pressure sodium lamp with IR reflective films

10. E. Kauer , Philips Tech. Rev. 26 - 2 (1965) 33
11. M.H.A. Weijer , ibid. 23 - 8/9 (1961/62) 246
12. R. Groth, E. Kauer , ibid. 26 - 4/5/6 (1965) 105
13. H.J.J. Boor, R. Groth , ibid. 29 - 1 (1968) 17
14. L. Sprengers et al , J. Illum. Engng. Soc. 14 - 2 (1985) 607
15. G. Blandenet et al , Thin Solid Films. 77 (1981) 81
16. R. Pommier et al , Thin Solid Films. 77 (1981) 91
17. L.V. Aleksandrov, A.S. Ivantsev , Applications of Surface Science 11/12 (1982) 563
18. A.S. Steven et al , Precious Metals (1981) 260
19. I.S. Goldstein et al , Lighting Research & Technology 18 - 2 (1986) 93
20. I.S. Goldstein , J. Light & Vis. Env. 9 - 2 (1985) 55
21. B.E. Yoldas , T.OKeefe , Applied Optics 23 - 20 (1984) 3638
22. K. Honda et al , Proc. SPIE (1983) 29
23. A. Kawakatsu et al , J. Illum. Engng. Inst. Jpn. 69 - 10 (1985) 537
24. L.E. Hoegler, T.K. McGowan , J. Illum. Engng. Soc. (1984)165
25. R. Martin, J. Rancourt , Proc. Annu. Tech. Conf. Soc. Vac. Coaters 29th (1986) 53
26. J. Rancourt et al , Proc. SPIE (1986) 653
27. A. Ohara et al , Acta Horticulturae 230 (1988) 349

Table. 1 Constitution and coating methods of IR reflective films

Type	Constitution		Coating method
Metal film		Au, Ag, Cu, Al	Vacuum deposition Sputtering
Transparent Conductive film		In_2O_3:F SnO_2:F In_2O_3:Sn SnO_2 :Sb Cd_2SnO_4	Chemical Vapor deposition spraying Vacuum deposition
Dielectric interference film	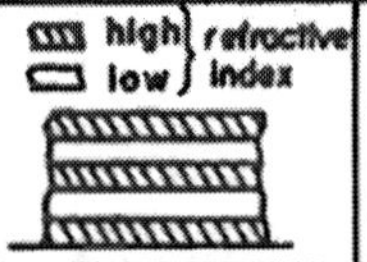	TiO_2-SiO_2 multilayer Ta_2O_5-SiO_2 multilayer ZnS-MgF_2 multilayer TiO_2-MgF_2-Ge-MgF_2	Vacuum deposition Dipping
Metal-dielectric film	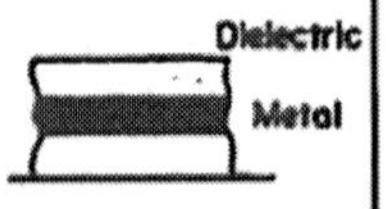	TiO_2 -Ag-TiO_2 ZnS-Ag -ZnS	Vacuum deposition Sputtering

Dynamics of the Neutral and Ionized Barium Atoms in the Vicinity of Fluorescent Lamp Electrodes

A. K. Bhattacharya and A. S. Awadallah
GE Lighting, Nela Park,
Cleveland, Oh, 44112, USA.

ABSTRACT

The useful life of a fluorescent lamp electrode depends upon the existence of a Ba-Sr-Ca oxide coating on the surface of the tungsten coil. During lamp operation the emission material is gradually lost by evaporation and possibly also due to sputtering from ion bombardment. Laser induced fluorescence techniques have been developed and utilized to measure the number density of neutral and ionized barium in the vicinity of such an electrode.[1,2] A theoretical analysis for a three level atomic system irradiated by a pulsed laser beam and results of LIF measurements of Ba atoms liberated from a F-lamp electrode are presented. Results of such measurements performed on electrodes of operating lamps reveal some very interesting insight to the loss processes for the emission mix: (1) The loss of emission mix is predominantly by thermal evaporation and the temperature of the electrode does not vary with the ac-cycle due to large thermal mass, (2) Neutral barium is thermally liberated from the electrode at a constant rate over the whole ac-cycle, and the measured variation in the neutral and ionized Ba signals are due to the varying degree of ionization of Ba in the electrode fall regions of the electrode and (3) The loss of Ba from the electrode, both as neutral and ionized atoms, occurs mainly in the positive half-cycle when the electrode acts as an anode.

Neutral Ba leaving the electrode surface is ionized in the high cathode fall (10–20 V) and are swept back on the negative (cathode) electrode. This results in low neutral and ionized Ba populations in the vicinity of such an electrode in the cathode half-cycle. On the other hand , when the electrode acts as an anode, relatively a fewer neutrals are ionized in the lower anode fall (5-12 V) and consequently a higher number of neutral Ba is detected. Also, the neighboring electrode is now at a positive (anode) potential and thus no sweeping action on the positive ions take place. This also results in a higher ion density near the anode.

The proposed mechanism is supported by results obtained from lamps operating on both dc and ac operations, where the voltage across the lamp was short-circuited for short time at various phases of the lamp current. Both neutral and ion signals increased to a fairly high values when the lamp voltage was removed for a period of 1–2 ms in the cathode half-cycle. Application of a negative potential (5 – 40 V dc) to the cathode lowered the ion signal significantly. Similar behavior for neutral and ions are observed for the anode half-cycle.

A rather small LIF signal for neutral Sr was detected near the electrode of a lamp operating at the rated lamp current. It showed behavior similar to that of Ba.

References:

1. A. K. Bhattacharya J. Appl. Phys., June, (1989).

2. A. K. Bhattacharya J. Appl. Phys., June, (1989).

THE VAPORIZATION BEHAVIOUR OF PURE BaO AND ALKALINE EARTH CONTAINING MIXED OXIDES.

L.Bencze, A.Popovič[x] and O.Kaposi
L.Eötvös University, Department of Physical Chemistry, H-1088 Budapest, Puskin u. 11-13, Hungary
[x]J.Stefan Institute and Department of Chemistry, University of Ljubljana, 61111 Ljubljana, Yugoslavia

Refractory mixed oxides like Ba_2CaWO_6, Ba_2SrWO_6, $Ba_3Y_2WO_9$ etc. are commonly used as cathode thermionic emission materials for the production of high-pressure sodium /HPS/ and high-pressure mercury /HPM/ lamps. The emission materials which coat the tungsten electrode coils vaporize by a dissociative process at high temperature according to the following type of reaction:

$$2\ Ba_2CaWO_6 + W = 3\ /Ba_{1/3}Ca_{2/3}/WO_4 + 3\ Ba$$

The formation of Ba as thermal decomposition product suggests that the electrodes can be activated by the thin Ba film on the surface of tungsten resulting an improved thermionic emission capability.

For the studying of dissociative evaporation process of cathode materials Knudsen cell mass spectrometry has been applied. The decomposition temperature, the dominant decomposition products and their sublimation enthalpy were determined.

Langmuir free vaporization and mass spectrometric surface ionization technique was used for the investigation of thermionic positive ion emission of cathode materials. The threshold temperature of the positive ion emission and the activation energy have been determined.

The fragmentation pattern of gaseous species over solid Ba_2CaWO_6/s/ at 30 eV ionizing electron energy and at 1700 K can be seen in Table 1.

Ion	Relative Intensity
$BaWO_4^+$	100
BaO^+	87,9
Ba^+	80,7
Ca^+	30,4
$BaWO_3^+$	11,4
WO_2^+	7,1
WO_3^+	5,0
$BaWO_2^+$	5,4

On the basis of ionization efficiency curves and temperature dependence of ion currents /Clausius-Clapeyron plots/ it was possible to find the precursor molecules of the ions. By means of the usual mass spectrometric calibration procedure the partial pressures and the enthalpy changes of sublimation of the vapour components, the equilibrium constants and the enthalpy changes of the following reactions were calculated:

$Ba_2CaWO_6/s/ \rightleftharpoons CaO/g/ + BaO/g/ + BaWO_4\ /g/$

$BaWO_4/g/ \rightleftharpoons BaO/g/ + WO_3/g/$

The ionization efficiency curves of the detected ions were evaluated by the very reliable EDD /electron energy distribution difference/ method of Winters at al. [1].

The enthalpy change of sublimation of BaO/s/$(\Delta_{sub}H^o_{1504K})$ in pure BaO was 397 ± 21 kJ.mol^{-1} which is in very close agreement with the previous literature values, $\Delta_{sub}H^o_0 =$ $= 439$ kJ.mol^{-1} [2], $\Delta_{sub}H^o_0 = 435$ kJ.mol^{-1} [3].

[1] Winters R.E., Collius J.H., Courchène W.I., J.Chem.Phys. 45, 1931 /1966/

[2] K.Hilpert, H.Gerads., High Temp. Sci. 7, 11-19 /1975/.

[3] M.G.Inghram, W.A.Chupka, R.F.Porter, J.Chem. Phys., Vol. 23, No. 11, 2159 /1955/.

Simulation of the Fluorescent Lamp Cathode

A. K. Bhattacharya, J. H. Ingold, T. F. Soules

GE Lighting, Cleveland, OH 44112

23 January 1989

Introduction. The temperature of the fluorescent lamp cathode must be high enough to provide lamp current of a few tenths of an ampere by thermionic emission, yet low enough to limit loss of barium oxide to an acceptible rate. In general, the temperature distribution along a cathode is highly non-uniform during normal operation. The highest temperature along the cathode occurs on the bare tungsten coil next to the emission mix in rapid start lamps, and on the emission mix in instant start lamps. Because the consumption rate of barium oxide is strongly dependent on temperature, the temperature profile is an important design consideration. The purpose of this work is to describe a thermal model of the fluorescent lamp cathode, which can be used to shorten the design cycle of new lamp types.

Method. The electrode assembly—lead wires and tungsten coil partially filled with emission mix—is represented by a segmented cylinder, with each segment divided into finite-difference elements. Average material properties of each element are based on wire length and diameter, and amount of emission mix in that element. The axial temperature distribution along the cylinder is calculated by solving a one-dimensional energy balance equation which accounts for the most important heating and cooling processes taking place. The effective gray body emissivity of the emission mix is found by comparing calculated temperatures with those measured for the case of DC auxiliary heat, but no lamp current.

A key assumption is that the ratio of ion current density to electron current density is proportional to the cathode fall voltage,[1] which is assumed axially uniform along the cathode. A second key assumption is that cathode work function is equal to the lowest patch work function, because calculated electric field at the cathode surface is sufficient to saturate the anomalous Schottky effect.[2]

Results. The most important heating mechanisms for rapid start operation are found to be Joule heating in the primary coil, plus electron bombardment during the anode half-cycle. For instant start operation, Joule heating in the primary coil, plus ion bombardment during the cathode half-cycle, plus electron bombardment during the anode half-cycle, are most important. The most important cooling mechanisms are found to be gray body radiation and electron emission during the cathode half-cycle. With no constraint on spot size, the model shows the formation and behavior of the hot spot with and without auxiliary cathode heat. The minimum energy principle is not required to give unique solutions.

Figure 1 shows calculated temperatures compared with those measured for seasoned instant start lamps, which have no auxiliary cathode heat, and for seasoned rapid start lamps, which have auxiliary heat applied simultaneously with AC lamp current. The model predicts a hot spot on the bare tungsten coil for rapid start lamps, and on the emission mix for instant start lamps, in agreement with observation. Based on this qualitative agreement, it is concluded that ion bombardment during normal operation is an important heat source for instant start lamps, but not for rapid start lamps.

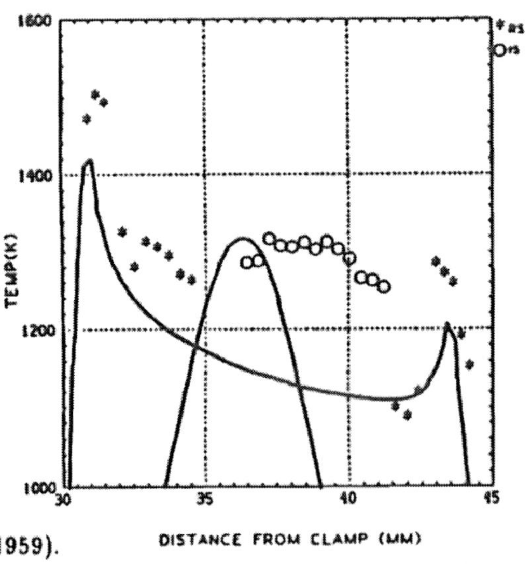

Figure 1. Comparison of calculated (solid curves) and measured (symbols) temperature along a standard fluorescent lamp cathode. Symbols: O-instant start; *-rapid start.

[1] J. F. Waymouth, *J. Appl. Phys.* **30**, 1404 (1959).

[2] R. H. Springer, private communication, 1988.

Thallium pressure determination from the Tl-377.6 nm resonance line in high pressure mercury + thallium iodide discharges

S.Couris, E.Drakakis and M.Aubes
Laboratoire de Décharges dans les Gaz (URA-277 CNRS)
118, rte de Narbonne, 31062 Toulouse, FRANCE

The knowledge of additive partial pressure is essential for the description of the emitted spectrum from metal halide discharges. The most common method used is that of the total intensity measurement of an optically thin line of the additive, the temperature profile is supposed to be known.

In this work the partial pressure of the additive is determined from the experimental contour of the Tl-377.6 nm resonance line emitted along a line of sight, knowing the temperature profile and the mercury pressure. The method proposed has been applied to two mercury-thallium iodide discharges operated at 50 Hz ac for which the corresponding mercury pressures were 1.7 and 2.5 bars.

The optical depth at the red maximum of the 377.6 nm resonance thallium line may be written as (1):

$$\tau_R = \frac{\lambda_o^4}{8\pi c} A_{ij} \frac{g_i}{g_j} \int_{-R}^{+R} n_j(x) P(\Delta\lambda_R;x) dx \qquad (1)$$

where : $P(\Delta\lambda_R;x) = \frac{1}{2}\frac{\Delta\lambda_o^{1/2}(x)}{\Delta\lambda_R^{3/2}(x)}$ and $\Delta\lambda_o(x) = \frac{\lambda_o^2}{2\pi c}\left\{\frac{4\pi}{3}\right\}^2 C_6\, n_{Hg}^2(x)$.

In our experimental conditions the dominant line broadening process in the red line wing is the quasi-static effect of mercury-thallium atom interactions for a Van der Waals interatomic potential (in the nearest neighbour approximation). Using Boltzmann's and ideal gaz law the expression (1) may be rewrite as follows:

$$\tau_R = \frac{\lambda^5}{6c\sqrt{2\pi c}} A_{ij} \frac{g_i}{k^2} C_6^{1/2} \frac{P_{Tl} P_{Hg}}{\Delta\lambda_R^{3/2}} R \int_0^1 \frac{dr}{U(T)T^2(r)} \qquad (2)$$

where: P_{Tl} is the partial pressure of the emitting element,

$U(T)$ is the partition function of Thallium,

C_6 is constant of the interaction for the Van der Waals potential with a value $2.4\times10^{-42}\ m^6s^{-1}$ (2),

A_{ij} is the transition probability for the 377.6nm Tl line taken $6.22\times10^7\ s^{-1}$ (3)

$\Delta\lambda_R$ is the distance of the red maximum from the central frequency.

Calculations of the optical depth at the red maximum of the 377.6 nm line, according to Bartels (4) approach, for different arbitrary central temperatures and thallium pressures have been carried out. An almost constant value of 2.06 for τ_R with a small fluctuation of about 2% has been obtained.

With these results one may finally calculate the partial pressure of the emitting element, using the formula :

$$P_{Tl} = \tau_R \frac{1}{\frac{\lambda^5}{6c\sqrt{2\pi c}} A_{ij} \frac{g_i}{k^2} C_6^{1/2}} \frac{\Delta\lambda_R^{3/2}}{RP_{Hg} \int_0^1 \frac{dr}{U(T)T^2(r)}} \quad (3)$$

In the above calculations the temperature profiles have been determined from measurements of the optically thin Hg-577nm line and from the 404.7, 435.8 and 546.1 nm self-reversed mercury lines. Simultaneously thallium partial pressures have been determined from measurements of the total intensity of the 655nm line. Table 1 shows that the two sets of results are in satisfactory agreement.

Table 1

t (ms)	$\Delta\lambda_R$ (Å)	$W\times10^{-8}$ (K^{-2})	p_{Hg} (bar)	p_{377} (mbar)	p_{655} (mbar)
0	2.32	4.63	1.7	9.2	10.1
5	2.16	3.80	1.9	9.1	9.1
0	4.6	5.05	1.5	24.5	31.7
5	4.4	4.41	1.7	24.9	28.9
0	7.4	4.59	5.2	17.7	16.1
5	7.1	4.50	5.2	16.8	14.9
0	11.2	5.30	4.8	24.5	31.7
5	10.6	4.92	5.0	23.9	24.1

(1) Drakakis, E., Thesis 3ème cycle, Université Paul Sabatier, Toulouse, 1988.
(2) Solyanikova, V.A., Opt. Spectrosc., 44, 136, 1976.
(3) Gallagher, A., Luvio, A., Phys. Rev., 136, 87, 1964.
(4) Bartels, H., Teil. Zeit. Physik, 128, 546, 1950.

S.Couris is indebted to *European Economic Communities* for partial support.

Electrode Temperature Measurements in Gas Discharge Lamps

E. H. Curtis and A. G. van der Kooi
Philips Lighting B. V., Eindhoven, The Netherlands

The determination of electrode temperatures in operating gas discharge lamps gives insight into many electrode functions including electrode energy loss, emission characteristics and rates of evaporation, sputtering and chemical processes. An Optical Multichannel Analyzer is given the special application of radiation profile detection of the subject electrode. Using this device, time and space resolved electrode temperature is measured via two-wavelength pyrometry. An image of the electrode is focused on the pixel array and through the use of superior quality optical filters at carefully chosen wavelengths, two wavelength are recorded. Two-wavelength pyrometry ensures for highly accurate temperature measurement as errors due to wall blackening, small variations of emissivity and image distortion are minimized.

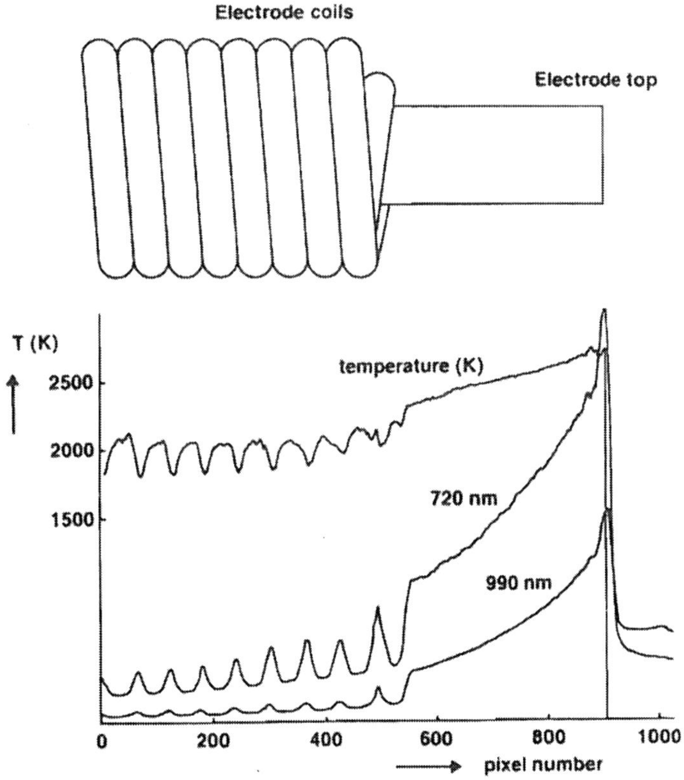

The measurement range, 1000 to 3500 Kelvin, is determined by the sensitivity of the OMA III detector and is calibrated via a tungsten ribbon lamp. Application of this method of temperature measurement is possible in all lamps types with visible electrodes. Shown in the figure are the two Planck radiation profiles and the temperature profile of an electrode in a metal halide lamp. Reflections between the electrode coils cause intensity peaks in the Planck radiation profiles. The accurate temperature measured is on the electrode coil surfaces which corresponds to the lows between the Planck radiation profile peaks and the highs in the temperature profile.

From a numerical fit of a region of the measured electrode temperature profile, a calculated electrode profile is extrapolated to the electrode top yielding the theoretical top temperature and electrode power input. The model used for this temperature fit is based on an infinitely long rod-electrode on which the arc is diffusely attached at the electrode top.[1] Through fast temperature measurement, time resolved measurements are performed and temperature variations in the anode and cathode phases are determined. Fourier manipulation of these 'ripple' files yields phase-resolved anode and cathode power. A second application of fast temperature measurement is in the determination of temperature behaviour during the glow-to-arc transition. After ignition, temperature and lamp current are recorded and show interesting results over the temperatures reached during the running-up process. The strengths of this technique lie in the accuracies possible and the flexibility of the OMA system. Future development in the area of temperature diagnosis of electric discharge lamps are certain.

1. Tielemans and Oostvogels, Philips Journal of Research, Vol. 38, Nos 4/5, 1983.

Evaporation Rates from an Oxide Coated Tungsten Electrode in Low Pressure Discharge Devices

Rex D. Groves and Ashok K. Bhattacharya GE Lighting,
Nela Park, Cleveland, Oh, 44112 USA

ABSTRACT

The tungsten electrodes of low pressure discharge devices, e.g., fluorescent lamps, are usually coated with oxides of Ba-Sr-Ca to enhance electron emission at low temperatures. The lifetime of such devices are primarily a function of the emission mix loss by evaporation. Each metal species reacts with the tungsten cathode to form tungstates and release the Ba-Sr-Ca atoms. This work uses resonance radiation absorption techniques to acquire absolute number densities of various species in the vicinity of such electrodes.† A simple diffusion model was used to relate such measured densities with the evaporation rates from the surface. These determined rates are compared with Raoult's Law.

Specially prepared samples of the carbonates (e.g. Ba/Ca, Ba/Sr) in molar weight percentages of 50/50 were used. As well, pure Ba and a triple carbonate of Ba/Sr/Ca in molar percentages of 39/31/30 were added. All were put onto electrodes and "activated" to reduce each sample to its respective oxide. After each were put in test cells of 2.4 torr argon, atomic absorption was used to determine the absolute number densities.

Each electrode was heated and the average temperature determined. A hollow cathode lamp was used to supply the necessary excitation for each species that was being observed. As the temperature was raised, species density increased exponentially in good agreement with theory. It was found that the heat of vaporization ΔH for Barium agreed quite well with previous work by Rittner.‡

Raoult's law predicts that the vapor pressure of a gaseous species should be proportional to the compositional molar fraction. It was uncertain whether this would be the dominant process that governed the species density above an electrode. Results of this work show that Raoult's law indeed is followed. In addition, the diffusion theory presented gives an estimate of the number density of the species and is compared with experiment. Work has been completed with Barium and agrees well with the projected values. Because of this, the model can give us a good estimate of the evaporation rate.

This work also investigates the effect of adding Zirconium to the carbonate before activation. Initial results indicate a reduction in evaporation rate by a factor between 2 and 3. A summary is given of the measured number densities.

† A.K. Bhattacharya. Bull. Am. Phys. Soc. 33, 138 (1988)

‡ E.S Rittner, Philips Res. Report 8, 184, (1953)

Barium Diagnostics in Front of Heated Fluorescent Lamp Electrodes

R. Hilbig, U. Chittka
Philips GmbH Forschungslaboratorium Aachen,
Weißhausstraße, D-5100 Aachen, F.R.G.

To lower the work function of the cathode fluorescent lamp electrodes are covered with an emitter material, which is a mixture of Ba, Sr and Ca oxides. The lamps fail when the emitter is consumed by evaporation and sputtering during lamp operation. For a hot electrode the emitter loss is mainly due to evaporation of barium. To get a measure of the consumption of the emitter material the atomic barium density distribution in front of heated electrodes of fluorescent lamps (TLD 36 W) has been studied by laser spectroscopy. This diagnostic method has been proposed recently [1].

The barium density profile has been measured by laser absorption techniques. A tunable cw-ring dye laser which is pumped by an Ar-ion laser is operated at the Ba resonance transition at 553.5 nm. The laser bandwidth (< 1 MHz) is much less than the Doppler width of the Ba transition of about 0.6 GHz . By tuning the laser frequency over the resonance line the absorption profile can be analyzed.

The high power of the dye laser light enables saturated absorption spectroscopy to get high local resolution of the density distribution. This technique has been used to study the radial density distribution. The laser light is split into an intense pump beam with $P > 20$ mW and a weak probe beam with $P < 0.01$ mW. Both beams are overlapped in front of the heated electrode (distance to the electrode: 1 mm; overlap volume: < 0.1 mm^3). The transmitted light of the probe beam is detected by a photodiode. As the pump beam is chopped, the photovoltage detected by a lock-in amplifier at the chopper frequency is a measure for the Ba density in the overlap volume. The radial density distribution has been studied also by measuring the profile of the chord integrated absorption and then calculating the radial density profile by Abel inversion.

Both methods deliver the same radial density distribution and also identical absolute values for the Ba density. As shown in fig. 1 the density on axis is $n_0 = 3 \times 10^{15}$ m^{-3} for an electrode temperature of $T = 1260$ K ($p_{Ba} = 2.8 \times 10^{-10}$ bar).

[1] *A.K. Bhattacharya, P.B. Hlahol: 4th Symposium on Light Sources, Karlsruhe 1986*

Investigations into the axial density profile are performed with laser powers low enough to avoid saturation (P < 0.01 mW). Here the laser light is sent through the lamp in front of the electrode (parallel or perpendicular to the main extent of the electrode).

The mean Ba densities are shown in figure 1 together with the result of model calculations as a function of the distance to the heated electrode (T = 1290 K). The Ba density distribution has been calculated in the following way: The diffusion equation has been transformed into a finite difference scheme. The corresponding linear system is solved by a successive overrelaxation method (SOR). The dimensions of the lamp and of the electrode have been used as boundary conditions. Cylindrical symmetry has been assumed. First the temperature profile has been calculated from the measured electrode temperature. Beside the measured density n_0 in front of the electrode this profile is an input parameter for the diffusion calculation.

The model delivers the axial and the radial density distribution of barium in front of the electrode. The axial density distribution as well as the radial distribution fit well to the measured profiles (see figure 1). Therefore, the calculated density profile could be used to determine the evaporation rate of atomic barium from the heated electrode. By integration of the barium flux through a surface around the electrode an evaporation rate of 0.0017 mg/h has been found (T = 1290 K).

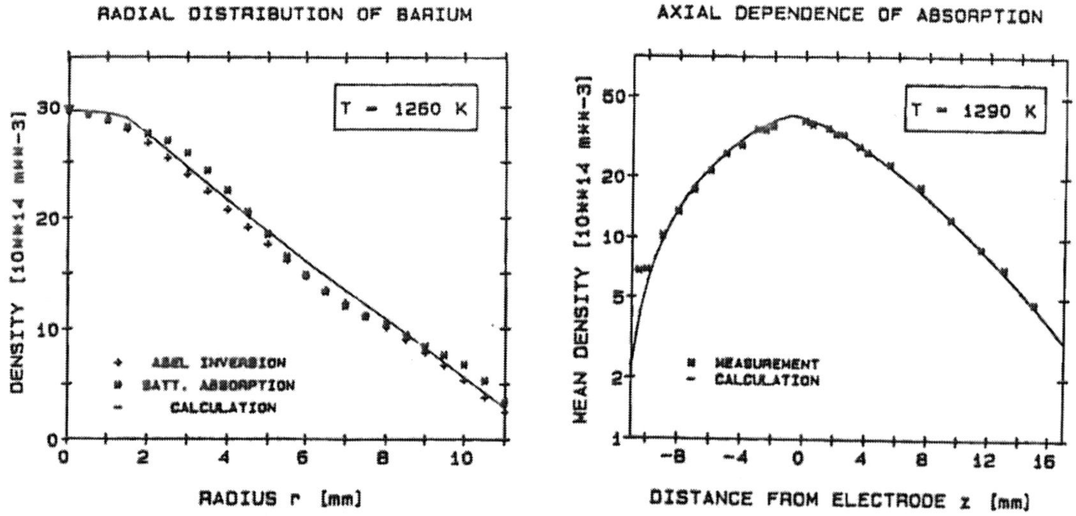

Fig. 1: Measured and calculated axial and radial density distribution of Ba near the heated electrode.

Space and time dependent modelling of glow and thermionic cathode spots

S. Holló, B. Nyíri

Tungsram Co. Ltd. Light Source Development Division
H-1340 Budapest, IV. Váci út 77. Hungary

For modelling of cathode spots we broke down the energy and charge transports at the electrode surface to lateral (tangential) and normal components. The lateral ones are treated spatially in differential equations, while the normal components are accounted for in the source terms of the balance equations (BE). So, spatial derivatives are meant here in 2 dimensions - parallel to the surface of the cathode.

Thus the BE for positive ions reads

$$\frac{\partial q}{\partial t} + div(\mu_i qE) = \frac{1}{1+\gamma}\left[\left(\gamma j + R\right)\cdot\eta(U,q) - j + R\right] \qquad (1)$$

Here q is the ion quantity in the sheath per unit area, E is the lateral electric field, γ is the secondary emission coefficient, $R(T)$ is the Richardson current density, $j = (1+\gamma)j_i + R(T)$ is the total current density. The total charge multiplication factor in the sheath, calculated from the one dimensional glow equations is

$$\eta(U,q) = \exp\left[2\varepsilon^2 \frac{U}{q^2} \int_0^{q/\varepsilon} \alpha(\xi)d\xi \right] - 1 \qquad (2)$$

with $\alpha(\xi)$ being the Townsend ionization coefficient. The cathode fall U is calculated with taking into account an external circuit with voltage supply and resistive ballast.

In the present approach the surface energy balance gets a

form of Fourier equation with nonlinear source term :

$$c\frac{\partial T}{\partial t} - \lambda \, \Delta T = \frac{j - R}{1 + \gamma} \left(K + V_i \right) - j \varphi - \Lambda (T - T_o) \qquad (3)$$

with T is the local temperature, V_i and K are the ionization and kinetic energy of the ions, φ is the work function and the last term accounts for heat transport to the bulk.

To obtain the numerical solutions of the above system of equations for the time evolution of T and q we modified Galerkin's method with that of Kantorovich.

The behaviour of the solutions is essentially determined by the roots of the right hand sides of the eqs. (1) and (3). The charge source has generally 1-3 roots and the heat source has 0-2 ones depending on the value of U in both cases.

Starting from various initial q and T distributions some well interpretable evolution phases can be separated:

- With quasi-homogene and low T, a small q perturbation collapses, while from a higher one a flat q distribution is developed. Here the current is approx. proportional to the spot area and U is constant. This accounts for normal glow.
- If the area of the surface is chosen smaller or the total current is increased, the previous case gives an increased current density with increased U. This is the anomal glow.
- In similar cases a small perturbation of T collapses while from a bigger one a sharp spot is formed. At high boundary T the final spot is wider and relatively smaller.
- At special circumstances from a developing glow-like q and T distributions a definite thermal spot evolves. This case is rather sensitive to the parameters of the equations.

TIME MODULATION OF SODIUM DENSITY IN A Hg-NaI AC ARC

D. Karabourniotis[+], A. Palladas[+*] and A. Tsakonas[+]

[+]Physics Department, University of Crete, 71409 Iraklion, Crete, Greece and [*]Centre de Physique Atomique, Universite P. Sabatier, 31062 Toulouse, France.

Recently a considerable effort has been made to determine the time-averaged density of neutral atoms in high-pressure mercury lamp arcs containing NaI as one of the additives *[1-3]*. Here we report experimental results concerning radial and time modulation of arc temperature and Na density in an axisymmetric Hg-NaI ac arc. The discharge tube, with 1.8 cm internal diameter and 4.5 interlectrode distance, was filling with 15mg Hg, 45 mg NaI and 40 Torr Argon. It was operated vertically from a resistively ballasted power supply at 50 Hz (U=63.5V, I=6.3 A), keeping constant the temperature at the lower part of the tube wall (735°C). Time-resolved emission measurements were conducted with a computer-automated system in a horizontal midplane of the arc. The optically thin Hg 577 nm line has been used to obtain temperature profiles (Fig. 1) and arc pressure (~1.5 atm). The optically thin Na 616 nm line has then been used (Fig. 2) to determine the Na density profile (Fig. 3) assuming LTE conditions.

We are indebted to Dr. Pabst (OSRAM - Munich) for the preparation of the experimental lamps.

[1] G.L. Rogoff et al, J. Appl. Phys. 62, 4084 (1987)
[2] J.T. Dakin and R.P. Gilliard, J. Appl. Phys. 60, 1281 (1986)
[3] K. Hilpert and J. Seehawer, TWAOG 12, 31 (1986)

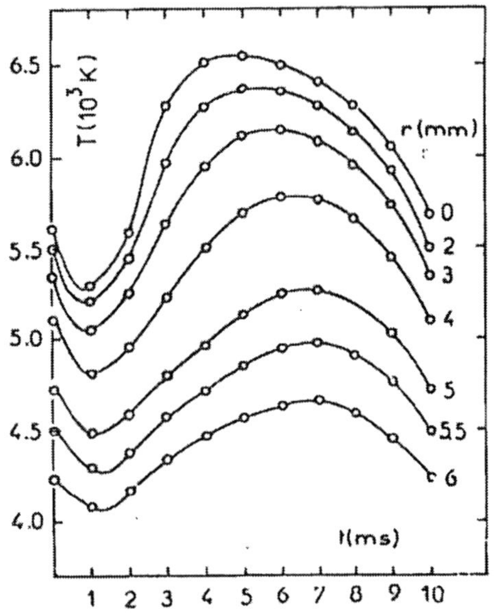

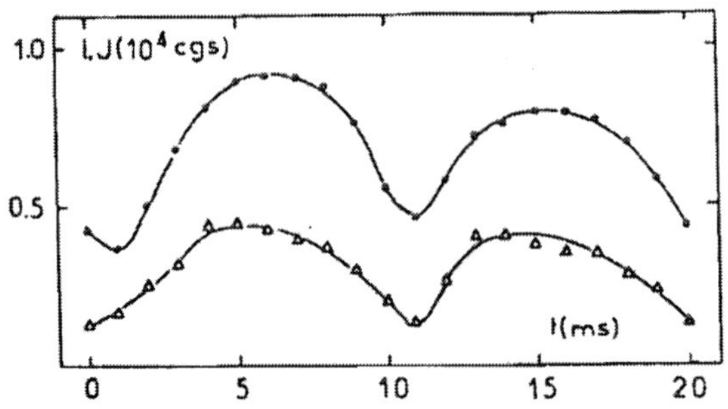

Fig. 2. Intensity I(0) through a diameter and emissivity J(Δ) at r=0 for Na 616 nm line.

Fig. 1. Temperature modulation

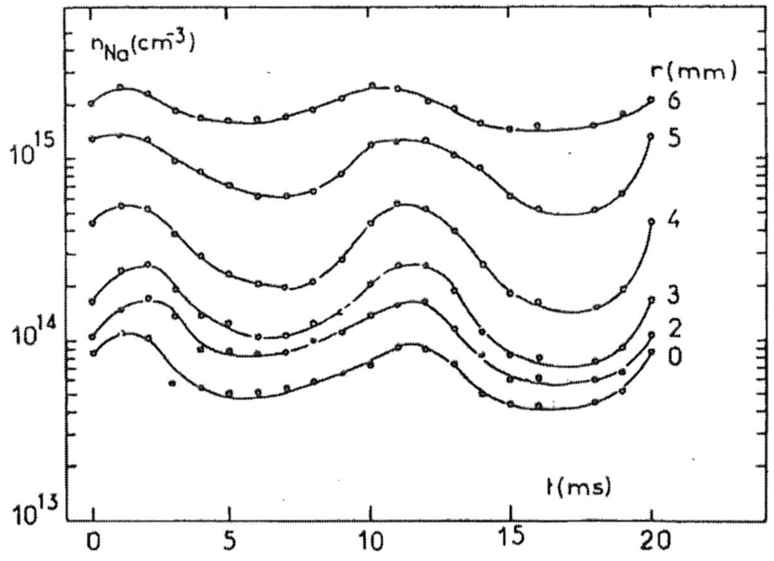

Fig. 3. Sodium density modulation

FUNCTION AND SURFACE PROPERTIES OF EMISSIVE COATINGS AND THEIR INFLUENCE ON HIGH PRESSURE GAS-DISCHARGE LAMPS

Ju.A.Korochkov, A.M.Kokinov, USSR

In view of energy-economic light source introduction the emissive coatings assume more and more significance that enables to gecrease near-electrode loss in discharge energy. The approach to solve a problem in choice of thermo-emitter composition for high pressure gas-discharge lamps is found with allowance for cathode work condition and refinement of thermoemitter function at all stages of lamp operation.

Having analysed work characteristic of cathode for high pressure gas-discharge lamps there stated the possibility for sufficient change in intensity of tungsten oxidative erosion due to surface temperature decrease with correctly selected thermoemitters in both starting and principal mode of light source operation. Performed refinement of emissive coating functions enabled to formulate requirements to thermoemitters those revealed dominating role of surface properties and substantiated the expediency of oxide compound use in gas-discharge light sources (1).

From comparison the features of starting and principal periods of lamp operation there stated the existence of dynamic and thrmodynamic equilibrium conditions in emissive coating function, that differ by component evaporation process. Theoretical analyse for vaporisation process of higher temperature oxides enabled to derive equations related component partial pressures to temprtature in indicated period of lamp operation. Based on theoretical and experimental study of thermoemissive property of high temperature oxide compounds there established the analitic dependence of work function coefficient on temperature and emissive centre concentration that enables to make estimation for emissive activity and atom evaporation rate in starting condition of discharge burning (1,2).

Experimental study of mercury and metal-halogen lamp operation identified the light output, color rendering index dependence on emissive coating composition. The principal reason of observed lamp characteristics is related to different intensities of erosion process, quantity estimation of which were carried out based on determination of tungsten mass contained in inner arc tube surface having burnt various lifetime.

REFERENCES

1. Korochkov Ju.A., Kokinov A.M., Rudnev V.N. Thermodinamic and emission properties of rare-earth oxides and peculiarities of its applications in discharge light sources.- 3 em Sumposium international science et technologie des sources de lumiere. Toulouse-France.-1983.-18-21 avril.

2. Korochkov Ju.A., Kokinov A.M. Thermoemission and evaporationof activated rare-earth metal oxides.-Proc.21st Session- Venice, 17-25 June 1987, Publication CIE n71(1987)-vol. 11-p130-131.

Estimation of convective arc bending by equivalent field method

By F. Márkus, B. Nyíri*, K. Gambár
Institute of Physics,Technical University, Budapest,
H-1521 Budafoki út 8., Hungary
*Tungsram Co.Ltd. Light Source Dev. Div. H-1340

The temperature distribution of a high pressure discharge lamp operating in gravitational field is influenced by the thermal convection; in horizontal operation the arc is bent. We calculated this influence for the case of weak convection. This task would require the simultaneous solution of the following, truncated, form of the heat balance equation and the Navier-Stokes equation:

$$\rho c\left(\mathbf{v}\nabla\right)T = \lambda\Delta T + \sigma\left(T \right) E^{2} \tag{1}$$

$$\nabla p = \rho_{o}\mathbf{g} - \alpha T\mathbf{g} + \eta\Delta\mathbf{v} \tag{2}$$

To simplify the calculations we applied the following approximations :

$$\sigma\left(T\right)E^{2}\lambda^{-1} = \begin{cases} 0 & \text{if} \quad T < T^{*} \\ A^{2}T+B & \text{if} \quad T \geq T^{*} \end{cases} \tag{3}$$

The anisotropy of the electric field distribution near the center of the tube can be a measure of the curvature of the arc. So by stopping at the first term in the expansion we get

$$E^{2} = E_{o}^{2}\left(1 - \frac{2\cdot x}{R_{o}} \right) \tag{4}$$

where R_{o} is the curvature to be calculated, and x is the vertical displacement. Thus, for given velocity and temperature distribution an estimate of R_{o}^{-1} can be obtained from the condition:

$$\int \left(\rho c (v\nabla) T \cdot \frac{1}{x} + \sigma(T) E_o^2 \right)^2 r \, dr = \min \qquad (5)$$

Consistently, calculating T(r) from (1) with (3) for the axisymmetric case, then with this solving (2) for v(r) eq.(5) gives for the curvature

$$R_o^{-1} = \left\{ \xi \left[\frac{\phi}{2} \xi^2 \left(\frac{1}{y} + \frac{1}{2}\right) - \frac{3\phi}{2} \xi \left(\frac{1}{y} + 1\right) + \frac{\phi}{3} \xi^3 \left(\frac{1}{2y} + \frac{1}{6}\right) + \left(\frac{43}{36} \phi - \frac{5}{6} T_o\right) \right] + \right.$$

$$\left. + x_o^2 \xi^2 \phi \left(\frac{1}{2y} - \frac{\xi}{16y} - \frac{\xi^2}{40y} \right) + \frac{5}{8} \xi^2 \left(\frac{\phi}{y} + T^* \right) \right\} \frac{2\rho c g \alpha R^2}{3\lambda\eta \left(1 - x_o^2/4\right)} \qquad (6)$$

Here x_o is the first root of $J_o(x)=0$, $y = x_o J_1(x_o)$, T_o is the wall temperature, R is the radius of the discharge tube,

$$\phi = \left(T^* - T_o \right) / \ln(\xi), \quad \text{and} \quad \xi = r^*/R \qquad (7)$$

and r^* is the radius of the arc that in the approximation (3), that is the radius of the cylinder where $T = T^*$.

The above method is not restricted for this simple case, since replacing T with the heat flux potential S ,and approximating ($\sigma(S)$ + radiation) in the form (3) our result (6) can directly be applied.

As our assumptions become inappropriate at high curvatures, when magnetic forces come into view, the inclusion of the latter is unnecessary. We think that an extension of this model to include the description of strongly bent arcs is of lesser importance, since the task is to control the lamp so as to keep the arc in the weakly bent state.

References: [1] *J.Verhás: Thermodynamics & Rheology.*

Technical Publishing House, Budapest, 1985.

ARC TEMPERATURE MEASUREMENT OF HIGH-PRESSURE DISCHARGES IN SCATTERING ENCLOSURES

S.A.Mucklejohn and A.T.Rowley
Lamp Research, THORN Lighting Limited, Light Sources Division,
Melton Road, Leicester LE4 7PD, U.K.

The measurement of arc temperature is one important diagnostic tool in the understanding of the complex chemical, radiative and heat transfer processes occuring within a high-pressure (HP) discharge lamp. This paper discusses the application of non-intrusive spectroscopic temperature measurement techniques to HP discharges contained within scattering enclosures.

Bartels' method [1] relates the maximum intensity of a self-reversed (optically thick) spectral line to the maximum temperature along a line of sight through the discharge. However, for many applications particularly HPS lamps, most optically thick lines are not self-reversed through the whole of the mains cycle. In addition, for typical conditions within a HP discharge lamp, the lines are only self-reversed in the very central region (the hottest part) of the arc. Moreover, the method relies on an absolute measurement of intensity along a line of sight through the discharge. This is not possible for arcs within scattering enclosures. It is possible to measure arc temperature from a relative Bartels' method using the relative intensity of two self-reversed lines.[2] However, there is some evidence to show that this is not applicable to scattering tubes.

It is demonstrated here that a method based on the relative intensities of two optically thin lines does work in such systems. It can be shown that the ratio of the intensities of two optically thin lines of the same element is:

$$\frac{I_\nu(1)}{I_\nu(2)} = \frac{\nu_1 A_1 g_u(1) \int_0^R \exp(-E_u(1)/kT(r))dr}{\nu_2 A_2 g_u(2) \int_0^R \exp(-E_u(2)/kT(r))dr} \quad (1)$$

where I_ν, ν, A, g, E and k have their usual significance and $T(r) = T_a(T_a - T_w)(r/R)^\alpha$ with T_a = centre arc temperature, T_w = wall temperature, R = arc tube radius and α = temperature profile factor. Such a method is shown: to be applicable to scattering tubes; to work for the whole of the mains cycle; to be valid for large volumes of the discharge and can be used on any species which produces at least two optically thin lines.

The principal sources of uncertainty in this method are the values of the transition probabilities (A_1, A_2) and the measurement of line intensities. The method can be checked on discharges in non-scattering tubes in regions where Bartels' method is known to apply; this gives a value for the ratio of transition probabilities in equation (1).

The method has been applied to metal halide lamps where the Hg (491.6, 577.0 nm) and Na (515.3, 615.4 nm) lines have been used and to HPS lamps where measurements of the Na 475.2 and 514.9 nm lines have, for the first time, given arc temperatures in an operating conventional HPS lamp.

References

[1] A.J.Davenport and K.C.Lapworth, National Physical Laboratory, Teddington, Middlesex U.K. Report Qu54 (1980) and references therein.

[2] G.Wesselink, D.de Mooy and M.J.C.van Gemert, J.Phys.D: Appl. Phys., 6, L27 (1973).

ZERO FIELD THERMIONIC EMISSION IN HF OPERATED FLUORESCENT LAMPS

H.J.W. Schenkelaars, J.J. Heuvelmans, A.v.d. Kooi,
J. Schlejen.
Philips Lighting B.V., Eindhoven, The Netherlands.

Lately, the interest for the electrodes in low pressure mercury fluorescent lamps has been renewed by the growing number of lamps operated on high frequency. The function of the electrode in a discharge lamp is to emit electrons in the cathode phase and to collect electrons in the anode phase. The electron emission capacity of a cathode can be determined by measuring the Zero Field Thermionic Emission (ZFTE). At DC and 50/60Hz the ZFTE is determined by measuring the amount of current the electrode can deliver in absence of the cathode fall. Small lamp currents have to be used, otherwise the necessary heating current becomes very high.

At HF operation the cathode fall behaviour is different from DC and 50/60 Hz operation. Phase resolved argon emission measurements have been performed in front of the electrode together with capacitive measurements of the AC cathode fall. Shortly after current reversal no cathode fall and no argon emission are present. The electrons are emitted thermally. At increasing current a very fast transition from thermal to thermionic emission occurs. The transition is characterized by a strong increase of the cathode fall together with a strong increase of argon emission (see figure).

The lamp is operated on a half bridge switch mode power supply (28 kHz). The thermal to thermionic transition is visible as a 'step' in the lamp current (due to the strong increase of the cathode fall). The lamp current at the 'step' is the maximium Zero Field Thermionic Emission. The main advantage of this method is that ZFTE can be determined in a very simple way without any influence on the lamp or on the electrode.

The I-ZFTE values of different electrodes deviate strongly. Hot spot temperature measurements have been done to interpret the deviation. The electrodes with the lower I-ZFTE appear to have the higher spot temperatures. This is explained by the higher workfunction at lower ZFTE.

The I-ZFTE determination can be used to investigate the influence of chemical (eg. emitter composition, contamination, processing) and physical properties (eg. electrode geometry, additional heating) on the electron emissivity of the electrodes in HF operated fluorescent lamps.

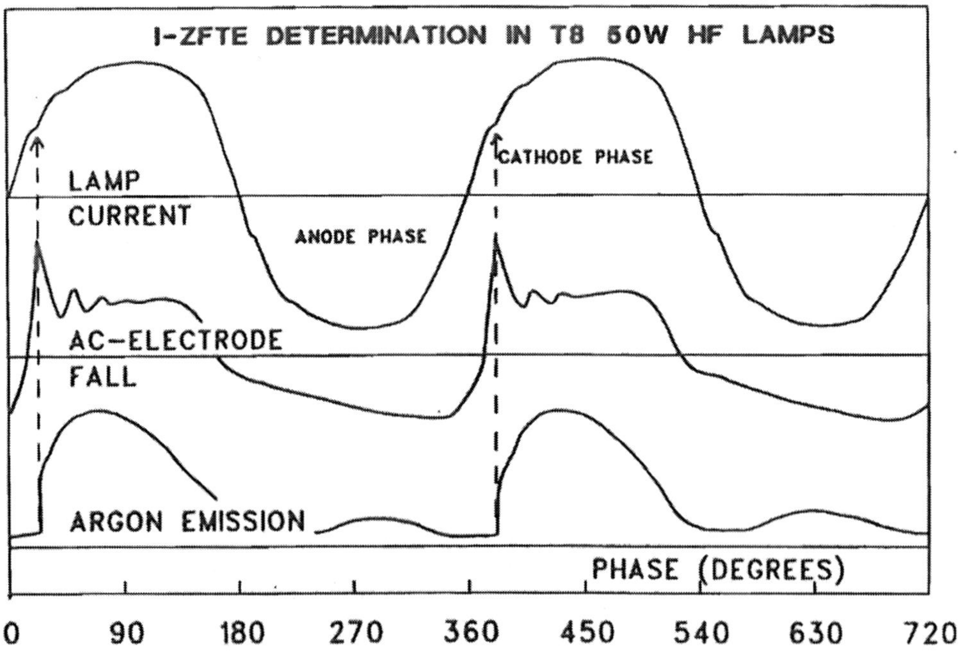

EXPERIMENTAL DIAGNOSTICS AND PHYSICAL CONDITIONS IN DISCHARGE PLASMA OF METAL-HALOGEN LAMPS

V.G.Vdovin, N.A.Vdovina, A.A.Pustoshkin, USSR

For edequate MHL modelling and searching ways for light efficiency increase it is necessary to carry out discharge diagnostics using methods, free of assumptions on local equilibrium in plasma. This requirement is not always taken into account. In our investigations /1-3/ we proceed from the most common nonequilibrium conception condition in plasma, assuming the ability of essential light efficiency increase.

By example of MHL (NaI, TlI, InI, Hg) there revealed nonequilibrium condition in nearaxial discharge region, with temperature gradient between the upper and lower arc tube parts 50°C over the range of practical iodide dosage, 3-10A, 50Hz, not using local equilibrium estimations, but determining microparameters experimentally, by means of analysis line contour width, shift and shape in a thin layer, holographic interferometry and other methods /1,2/ those not requiring apriory assumptions. As shown in figure, heavy particle temperature T_a in MHL is appreciably lower the electron temperature T_e, estimated within the limits of nonequilibrium theory (MDA) /3/ by measured nonequilibrium level population of atoms Na, Tl, In, Hg, at all binding energy ranges with an account of conversion reactions of molecular ions and associative ionisation. In this case the excitation temperature of first resonance additive levels in phase of MHL maximum current attains T_e value and may be laid into basis of electron temperature estimation. The analysis of plasma decay after current shunting enabled to define a relaxation character of level population, recombination flow ~(2 ÷ 5) x x 10^{18} cm^{-3}c^{-1} and to explain the reaction role

$$M^+ + 2Hg \rightleftarrows (MHg^+)\,Hg,\ (MHg^+) + e \rightleftarrows M^* + Hg.$$

in high excitation level impoverishment M- (Tl, Na, In, Hg).

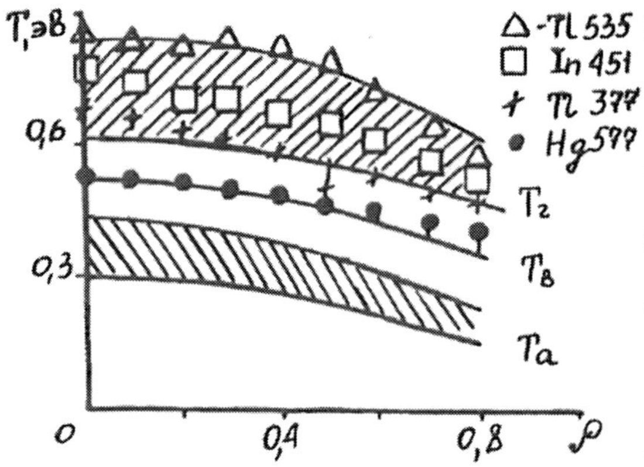

The results allow to assume: incorporation of additives doesn't result in decrease of mercury discharge temperature T_e due to associative ionisation with the participation of Hg buffer gas atoms, but can result in T_a decrease. With conceptions obtained there realised MHL with (NaI, ScI, Hg) and (NaI, TlI, Hg, Xe) P = 400, 700, 1000W ; 120 - 135 lm/W without load increase on fused silica wall.

References

1. V.G.Vdovin, A.D.Hahaev. ICPIG-14, Grenoble, France, Journal de physique, 1979, C7, t40, p.177-178.
2. V.G.Vdovin, N.A.Vdovina. ICPIG-14, Grenoble, France, Journal de physique, 1979, C7, t40, p. 839-840.
3. V.G.Vdovin, N.A.Vdovina, O.V.Shuvatova. ICPIG-15, Minsk, USSR,1981, Contr. Pap., partII, P-1616, p.941-942.

SPECTRA AND DIAGNOSTICS OF HIGH PRESSURE SODIUM-CADMIUM LAMP

Damir VEZA, Jadranka RUKAVINA#,
Davorka AZINOVIC and Goran PICHLER

Institute of Physics of the University,
POB 304, 41001 Zagreb - Yugoslavia

#Electric Lamp Factory,
Folnegoviceva bb, 41001 Zagreb - Yugoslavia

Recent spectroscopic measurements of high-pressure sodium-cadmium lamp spectrum ([1],[2]), pointed out need for basic spectroscopic data for this particular discharge system. Measurements of temperature and electron density are performed with burner made of polycristalline alumina operated on a 50 Hz AC power supply, in the range of lamp loadings (AC voltages) between 230W-650W (60V-140V). Using box-car averager we measured time-resolved spectra in order to find time evolution of sinusoidally driven NaCd plasmas during one current cycle. Figure 1 shows dependence of the electron density upon the overall lamp voltage (a), and upon the phase of the driving current (b), where "0°" represents minimum and "90°" maximum current. Electron density has been estimated on the basis of Inglis-Teller relation [3], using atomic lines from P-D, S-P and P-P series in sodium atom, and points in Fig.1 represent an average of these data. Compared to the analogous NaHg HP lamp this system shows about 30% lower electron density which does not saturate with lamp loading. We have not definite answer to the differencies between NaCd and NaHg discharges, but lower cadmium activity in the NaCd amalgam and more uniform (flat) radial temperature distribution instead of parabolic dependence may be a reason.

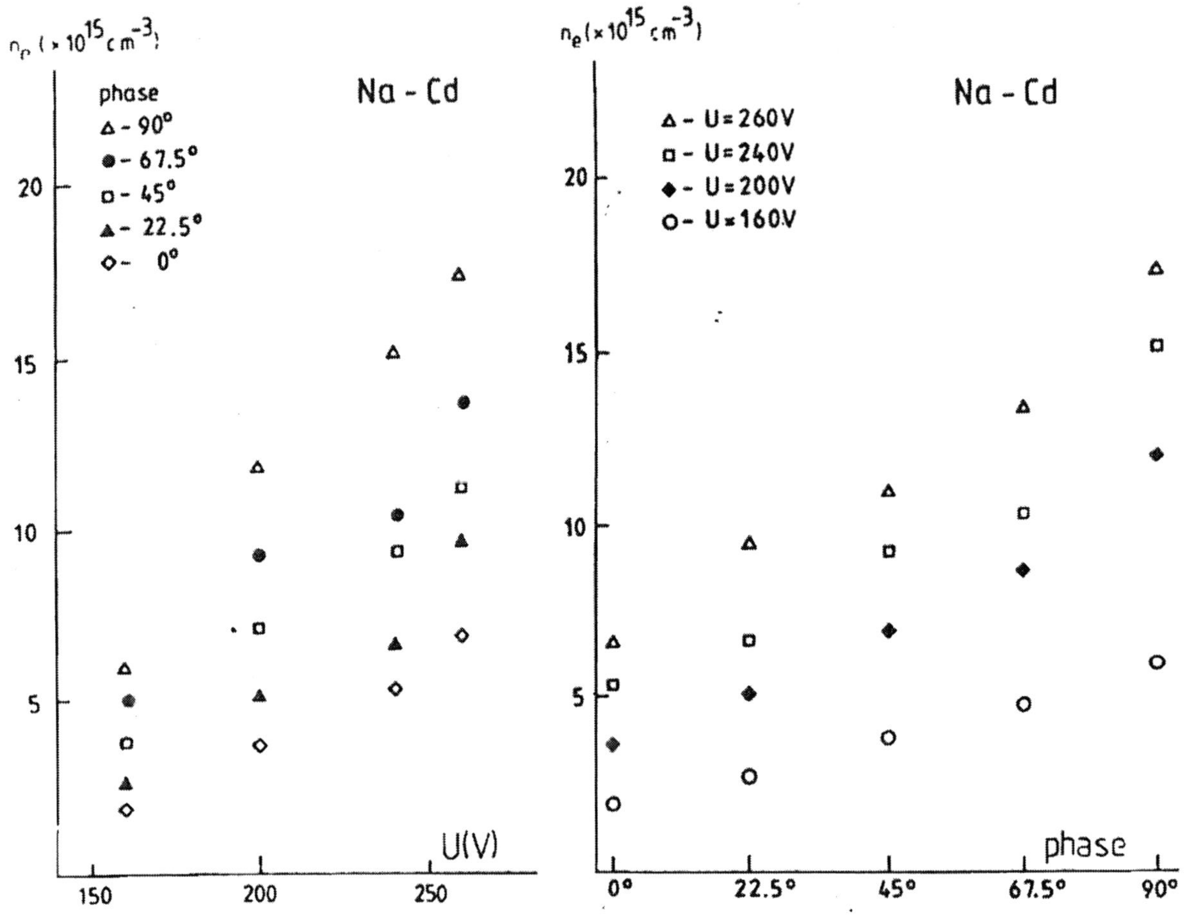

Fig.1. Electron density vs. voltage (a), and current phase (b)

Electron temperature has been measured via the shape of ion-electron recombination continuum to the sodium resonance level(s) [4]. Fortunately, in the NaCd discharge this part of spectrum is unaffected by other continua [1], and this method may be used with confidence. We found that temperature varies sinusoidally over a cycle, swinging between 2000°K and 4300°K, with minimum and maximum values depending on the lamp load. Continuation of these experiments using clear sapphire tube and laser spectroscopy techniques is under way [5].

REFERENCES:

1.G. Pichler, D. Veza and D.Fijan, Opt.Commun. 67 (1988) 45.
2.D.Fijan, D.Veza and G.Pichler,Chem.Phys.Lett. 154 (1989) 126.
3.H. Griem: *Plasma Spectroscopy*, McGraw-Hill, New York 1964.
4.D.E. Rothe, J.Q.S.R.T. 9 (1969) 49.
5.D. Veza, will be published.

Electrode Diagnostics Using Plasma-Enhanced Photoemission.

David R. Woodward and M. Bruce Schulman

Philips Lighting Company

330 Lynnway, Lynn, MA 01901, U.S.A.

A new laser-based technique is reported for non-intrusive in situ electrode diagnostics in fluorescent lamps [1]. The technique utilizes low-power pulsed laser beams of various wavelengths to probe the effective work function of the cathode in an operating fluorescent lamp. Plasma-enhanced photoemission (PEP) has been previously used as a diagnostic to monitor the effects of chemical changes on electrode surfaces due to RF plasma exposure [2-4].

Cathodes of operating fluorescent lamps were subjected to short pulses of monoenergetic photons, and photoemission caused electrons to be ejected from the surface, provided the photon energy was greater than the effective work function of the area illuminated. The corresponding burst of electron current and the resulting plasma-enhanced optogalvanic signal was monitored as a transient voltage change across the ballast resistor. This laser-induced photoemission current from the cathode depends on the local work function, temperature and accelerating fields at the surface.

The attenuated beam from the pulsed laser was directed perpendicularly onto the cathodes of the fluorescent lamps studied. The phosphor had been wiped from the ends of these lamps prior to processing to allow a clear path for the laser beam, which was focussed to a diameter of less than 1 mm. The lamp fixture was mounted on a precision translation stage so that the laser beam could be precisely positioned along the length of the cathode. The transient voltage pulse across the ballast resistor beginning with the arrival of the laser-beam pulse was measured by capacitive coupling into the 50-ohm input of a digital storage oscilloscope. No measurable signal was observed if the laser beam was moved just off the electrode into the surrounding plasma. This showed that the observed signals were due to plasma-enhanced photoemission from the electrode and not to a gas-phase optogalvanic effect.

The photoemission signal in 40W argon-filled and 34W krypton-filled fluorescent lamps was measured as a function of illuminated position on the cathode. The threshold photon energy for which photoemission occured was measured at several positions. The dependence of the signal on laser power and heating current was also investigated, and cases in which the photoemission current was space-charge limited were identified. There was a significant difference in the time dependence of the plasma-amplified optogalvanic signal for argon- and krypton- filled lamps.

[1] D.R. Woodward and M.B. Schulman, to be published.

[2] S.W. Downey et al., J. Appl. Phys. 63, 5280, 1988.

[3] G.S. Selwyn et al., Appl. Phys. Lett. 52, 1953, 1988.

[4] G.S. Selwyn and J. Singh, to appear in IEEE J. Quant. Electronics.

Electrode Phenomena

F. Nágel

Tungsram Co Ltd. Light Source Development Dept.
Budapest, IV. Váci út 77. H-1340 Hungary

Electrode phenomena in lamps make a hopelessly disarranged pile of hard to understand facts. To order these one feels great temptation to find a smart theoretical framework either to force observations in it or neglect them. Thus, we shall neglect electrode phenomena in incandescent lamps, such as oscillations in vacuum lamps, ion emission induced arcing in gas filled lamps, and arcs at the end of the life of the tungsten coil of higher voltage GLS lamps.

With regard to high pressure discharge lamps, electrode phenomena are highly involved in all stages of discharge evolution: breakdown, transient effects after breakdown, heat and material transport phenomena, including electrode chemistry, in the stationary state of AC and DC lamps.

As in discharge lamps the arc is usually self sustainig, the behaviour of the electrode is governed by various feedback loops. The thermal interaction between the plasma and the bulk of the electrode is to be mentioned first, since the temperature of the active surface determines electron emission and the speed of chemical reactions. So it has a strong influence on that how feedback loops of chemical transports to/from the plasma and the electrode bulk operate. Naturally, besides the above examples there exist several other feedback mechanisms such as cathodic space charge - cathode fall - ionization rate, or electic field at the surface - field emission - space charge - electric field again.

As a first example of feedback mechanisms let us regard a simple one, an observation with a certain starting aid, which is a pin near the outer side of the discharge tube. It enhances electron avalanches by its strongly inhomogeneous field during the ignition of the lamp. Now, contrary to (our) expectations the effect of the pin prooved to be the greatest in the vicinity of the anode (and not the cathode). This can be explained as follows: placing the pin near the cathode the electrons move to the wall due to their higher mobility. So by charging the tube the field is shielded. Now, when the pin is located near the anode, then only the ions, which have a much smaller mobility, could shield the field. Thus in this case the electron avalanche has more time to grow, to ionize. This surface charging mechanism is essentially the same as that governs a low pressure discharge when it follows the curvature of the discharge tube.

Before turning to topics more closely related to everyday lamp electrode phenomena, let us examine an experimental situation that shows many characteristics of a real lamp electrode, but the thermal feedback loop is cut.

Taking a gas filled diode having an indirectly heated dispenser type cathode (activated with e.g. Ba_2SrWO_6 or Ba_2CaWO_6). In series with the anode there is a variable resistor, an ammeter, a switch, and a 300 V source. Parallel to the diode there is a voltmeter.

Let us start the experiment with say 1000° cathode temperature. Decreasing the resistance the current is increasing monotonously while the diode voltage remains practically constant (approx 10 V). At a definite current, say 2 amps, the current suddenly falls down to say 3mA (proportional to rate of surface diffusion and the emission material loss), the voltage jumps to 300 V and a green flash can be seen. Changing the resistance to the maximal (initial) value does not alter the situation. Switching off for a few seconds the system 'recovers'. A recovery can be obtained without switching on and off by heating the cathode to say 1200°. It is clear that the recovery is based on the surface diffusion of the activator to the emitting surface.

Operating a similar but vacuum diode with contaminated anode, from AC source, the current-voltage characteristic shows a hysteresis loop. During the back-contamination from the anode to the cathode the backward line of the loop (corresponding to decreasing current) runs lower than the forward line, while after the contamination is over the backward line becomes the higher. Thus, in the first case the cathode is desactivated by poisoning ions, and in the second case the activation by current becomes dominating.

Returning to self sustaining arcs let us have a look at cathode spots. The fundamental difference lays in the fact that the electrode surface temperature is not externally governed but is determined by interactions with the plasma surrounding the cathode.

If there were no processes supplying heat to the cathode surface, the electron emission - together with the backward heat conduction - would cool the cathode surface down. At lower surface temperature - at lower current density - the heating back from the plasma is also small, thus in a temperature range there is a negative feedback loop. Of course, over a current or surface temperature limit T^* - due to the back heating - this feedback becomes positive.

Now, if the surface temperature is everywhere below T^* the thermal spot extinguishes. However, if a (local) perturbation rises the temperature so that it exceeds this limit, then a sharp spot starts to develop there. Its final sharpness depends on the total current. On the positive feedback parts of the surface the temperature is limited by the lateral heat transport. At places, where the feedback is negative, the temperature is lower than that would correspond solely to lateral transports. This spot is, evidently, unstable to finite perturbations.

If the global surface temperature is above T^* there is a positive feedback everywhere, so we get a stable diffuse spot, controlled by the heat transport to the boundaries.

The behaviour of the glow can be discussed similarly. The main feedback loop consists of the ion induced secondary emission from the surface and the ionisation by the emitted electrons in the high electric field domain.

At low ion densities - weak field - we have a negative feedback, since the total ion-multiplication factor of the loop is less than 1. Over a limit in space charge, i.e. in ion density, the net ion source - the feedback - becomes positive. So, an ion density perturbation (e.g. by ignition) below this limit is extinguished.

A further increase in ion density increases the field, but the ionisation rate reaches a saturation value (B Nyíri, this volume), while at the same time the thickness of this high field intensity domain decreases, due to electrostatical reasons. Because of this, there is another critical ion density, over which the feedback becomes negative again. This gives a limit to current density too. So, the current density is nearly constant in the area of positive feedback - determined by the total current. At places where the feedback is negative, the current density and space charge fall faster than required by lateral transports. A more thorough, and more mathematical discussion of these can be found in the paper of S. Holló & B. Nyíri (this volume).

In a discharge between electrodes, charged particles are emitted from one electrode and are absorbed by the other. This is chemistry in the broader sense, as we know that oxidation

and reduction are, in principle, equal to electron transfer.
So, investigating any electrode phenomenon, we should clear that wether current flow means reduction on the cathode and oxidation on the anode in the conventional sense too, or not. When not - why, when yes - how ?

In spite of the non-exdstence of a correct theory of emissive materials, it seems to me obvious to use oxides as emitters in vacuo or inert gases. It is far from being so obvious to use these in sodium or metal halide arcs. But we do it, and often with success. In other, seemingly similar, but unsuccesful cases the emissive materials are the dominant oxigen sources of the lamp.

Several oxides, including the ones we use as emitters, are n-type electric conductors (as they are in a little bit reduced state). These oxides normally conduct current without chemical consequences, but over a definite current density an oxigen vacancy conduction begins, resulting in an electrolitic decomposition. This critical current density is not a material constant determined by the composition only, but it depends on the prehistory of the oxide, above all, on the circumstances at the formation of the oxide.

Surprisingly, when an oxide coated anode is bombarded by electrons of increasing energy, nothing happens (except the well known desorption); not even if the electron energy reaches or passes the energy region of the chemical bond. But, increasing the current density with electron energy held constant (say 50 to 300 V) the decomposition of the oxide suddenly begins when the current density passes a definit threshold. This threshold has negligible temperature dependence. In a system in which both electrodes can be heated but which are containing different emissive materials, turning the direction of the electron current we find that the critical current density is independent of the direction, and it is equal to the lower. In real lamps the current density is usually higher on the cathode or in the cathodic phase, so practically we can neglect the decomposition on the anode. When investigating emissive materials in model devices, it is often easier to determine the critical current density on the anode.

Investigating the warm up transient of high pressure discharge lamps Holló and coworkers (Proceedings of the Symposium on High Temperature Lamp Chemistry 1988. Atlanta pp. 272) found that significant emissive material loss occures not during high energy ion bombardment of the glow phase, but mainly during the quickly moving spot (QMS) phase. Another interesting property of the QMS is that though it usually follows the glow it can often be observed in the glow - steady arc spot - QMS sequence. Moreover, the QMS can be followed not only by steady arc spot but also by glow.

Now, on the basis of the above described observations and spot models we may attempt to explain the mechanisms governing the QMS, where locally a similar event takes place than in model diode passing critical current density. The locally increased current density cleans the surface at a spot, thus the emissivity falls there. The global state (temperature) of the electrode, however, allows the development of either a glow or a thermal spot at another place. We also have to remember that here the plasma composition is not the average one but is strongly influenced by the material coming from the electrode. This produces a strong similarity between vacuum arc and a starting lamp, in spite of fact that a lamp before starting is gas filled.

Electrode phenomena are influenced also by chemical processes that have originally nothing to do with that they happen to be on an electrode. For example, Me_3WO_6 type emitters (where Me can be Ba,Sr,Ca) are perovskites. They can be contaminated by Al_2O_3 . Fireing Me_3WO_6 several times with some Al_2O_3 the perovskite structure shall be step by step destroyed, Me-aluminates appear. The same happens when repeating the experiment with Al_2O_3 contaminated Ba-zirconate (which is a perovskite too). This excludes the explanation of the aluminate production that Al_2O_3 is less volatile than WO_3. The cause is complex building tendency of Al. But, from other experiments it is well known that basic aluminates are more reducible by tungsten then tungstates. This shows us that equilibrium calculations, neglecting the reaction speeds, can be seriously misleading, and that Al contamination can be an important cause of end-blackening in HPS and fluorescent lamps. During blackening an anomalous loss of emissive material and in HPS lamps an anomalous sodium loss occure.

Stabilisation of White SON lamps

R.A.J.Keijser, Philips Lighting BV, Eindhoven, The Netherlands

High pressure sodium lamps with improved colour rendition (CRI > 80) and a colour temperature much higher than for standard HPS lamps (T_C around 2500 K) are a promising alternative for incandescent lamps in indoor lighting applications.

That such lamp properties can be achieved by raising the sodium vapour pressure inside the lamp has been known for some considerable time [1]. It was not untill recently, however, that the problems associated with detrimental chemical reactions at these high sodium pressures became sufficiently under control to obtain adequate lamp life [2,3].

Apart from this, control of the lamp colour is of extreme importance. For this type of lamp the colour of the emitted light is directly related to the sodium and mercury pressures in the discharge tube under operating conditions. Colour control, therefore, implies control of the temperature and composition of the liquid amalgam.

Small variations in these quantities can easily arise as a result of e.g. production tolerances in the lamps, alterations in the lamp during life or variations in line voltage. A proper choice of the lamp circuitry can be of great help in keeping operating pressures in the lamp and lamp colour within narrow tolerance limits. This has lead to a new approach for this particular category of HPS lamps, which has opened the door for

application of such lamps in colour critical situations. Practice has shown that with a relatively simple control circuit and an adequate control of the relevant lamp design parameters a good colour quality can be guaranteed during lamp life, irrespective of the external conditions.
In this paper several aspects of this stabilisation problem will be discussed for the case of a 50 Watt White SON lamp. The effect of variations in the amalgam temperature is studied via a method in which argon gas is let into the lamp outer bulb. By varying the gas pressure the amalgam temperature can be modified in a controlled way and the consequences for the vapour pressures and lighting properties can be determined for various possible circuits. It appears to be possible to completely eliminate the influence of variations in the heat balance on the lamp colour with a control circuit based on $V_{la} + \beta I_{la}$ = constant. Furthermore the influence of the amalgam composition has been investigated: for practical circuits the Na content of the amalgam should be kept in between the limits $0.72 < x_{Na} < 0.82$. Finally, some dynamical aspects, related to the stability of these systems e.g in case of external disturbances, will be dealt with.

[1] Mizuno,H.,Akutsu,H. and Watarai,Y.,CIE XVII Session, Barcelone,vol.21B,p71.14(1971).

[2] Akutsu,H.,Watarai,Y.,Saito,N and Mizuno,H., J.Illum.Eng.Soc.,Vol.13, pp.341-349(1984).

[3] Van Kemenade,J.T.C.,Keijser,R.A.J. and Bolt,B.F., J.Illum.Eng.Soc.,Vol.16, pp.150-161(1987).

INSTANT LIGHT METAL HALIDE LAMP

R.H.Springer, R.L. Hansler and J.M. Davenport

G.E. Lighting, Nela Park, Cleveland, Ohio 44112

A major limitation of extending the metal halide lamp technology into the automotive field has been the inability to provide sufficient light during the start up stages of the lamp. A new concept (1,2) based on the optimization of the discharge using a combination of high pressure Xenon and metal halides has been proposed. However,the high Xenon pressure and relatively large starting currents result in system constraints that are not present in standard metal halide systems. The practical implementation of this concept requires the solving of several technical problems, such as, starting, instant restart, ballast programming, arc stability and electrode design. The high pressure starting and hot restart can be solved separately (2), but the electrode and ballast design require a system approach. One of the most difficult problems is to develop design principals for electrodes that will not melt during high current starting, but yet have a stable spot during low current operation.

The high current and high Xenon pressure needed for the instant light favors the initiation of a cold cathode arc mode, rather than the cathode glow mode, during starting. Within one microsecond after breakdown the electrode fall voltages are low and the current capability of the spot is very high, provided the arc current is maintained at a high enough level to maintain this mode until the electrode becomes a thermal emitter. This approach to starting matches and measured arc centerline of a 400 watt mercury lamp.

Secondly, based on the above result, a downward curved arctube with offset electrodes is chosen to perform a further study which investigates the degree of uniformity of the wall temperature distribution, and the impact of arctube inclination angle with respect to gravity on the transport process and the resulting temperature distribution. It is found that the predicted and measured temperature profiles along the top and bottom walls are in good agreement. For the horizontally mounted arctube, both the prediction and measurement show that the interactions of the curved high temperature core, resulting from the balance of the buoyancy and the electric field, and the bent arctube shape, gives the top wall temperature a double-peak distribution while the bottom wall temperature shows a single-peak distribution. Both photographs and numerical predictions suggest that the gross patterns of the gas temperature distribution in the mid-plane are rather insensitive to the inclination angle. The wall temperature distribution, however, exhibits substantially greater sensitivity to the inclination angle, changing the top wall temperature from a symmetric double-peak profile for horizontal mounting to progressively more asymmetric single-peak profiles with increasing inclination angle. This change of characteristics in wall temperature distribution has important effects on the performance of the arctube.

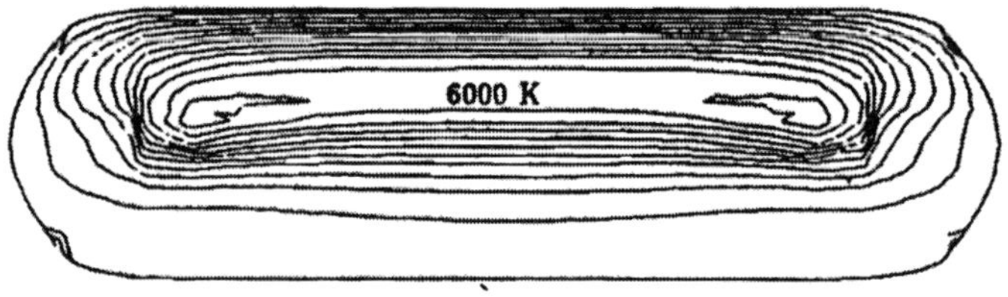

Fig.1 Temperature contours in central plane. (a straight arctube with centered electrodes)

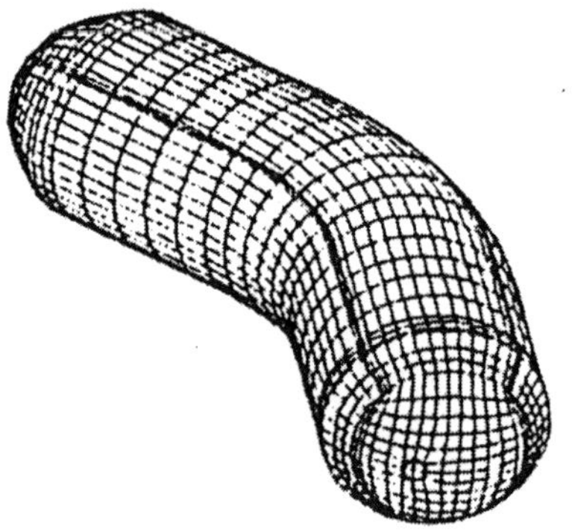

Fig.2 Computer generated geometry and mesh in 3-D perspective. (a curved arctube with offset electrodes)

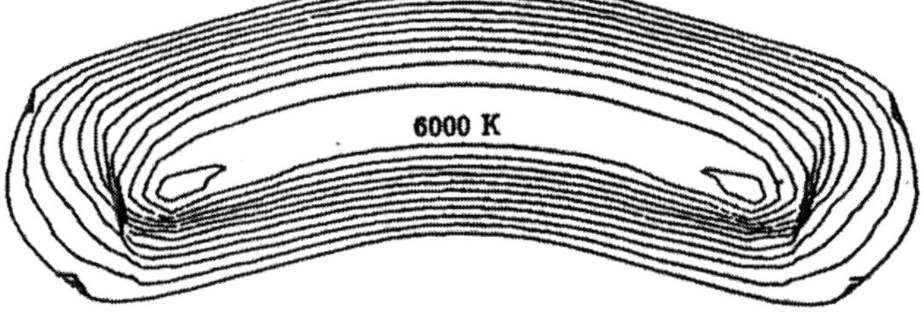

Fig.3 Temperature contours in central plane. (a curved arctube with offset electrodes)

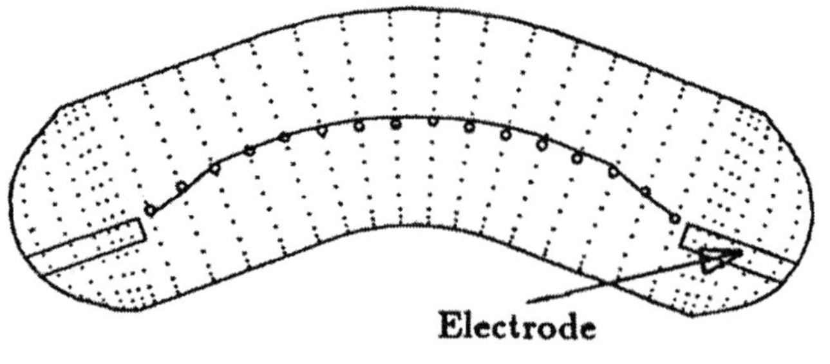

Fig.4 Arc centerline.
—— calculated result
o measured data

A COMPUTER MODEL OF A HIGH PRESSURE DISCHARGE LAMP BASED ON THE FLUENT CODE

L.Cifuentes and D.A.J.Mottram

Lamp Research, THORN Lighting Limited, Light Sources Division,

Melton Road, Leicester LE4 7PD, U.K.

S.Vasquez

Flow Simulation Limited, Sheffield, U.K.

Modelling a high pressure discharge lamp involves the simultaneous solution of the partial differential equations which describe fluid flow, heat transfer (including radiation) and electric field.

The development of such a piece of software is a lengthy business, as it involves the choice and implementation of the method of analysis (finite differences, finite elements, etc.) and, especially if allowances are made for a variety of geometries, boundary conditions, data formats and the like, plus the all-important pre-processing (data input) and post-processing (results output) facilities. The whole process can take several man-years.

Based on our experience in the development of a 1-D (radial) model of a high pressure discharge, it was decided that an advantageous way to tackle a 2 or 3-D computer model was to start from an established, commercially available software package which already implemented at least part of the required model. Several packages were evaluated [1] and a choice made.

The chosen package is called FLUENT. It implements a finite-differences approach to the solution of the fluid flow and heat transfer equations, including laminar and turbulent regimes, conduction, convection and radiation with conjugate heat transfer and offers a range of other facilities. It is fully interactive, runs under UNIX, VMS and MS-DOS on a range of machines from PC's to supercomputers and is very user-friendly.

Fluent, as it stood, was unable to model a discharge lamp, due to its lack of a solution of the current continuity equation, potential and current boundary conditions and radiation from the lamp walls. The missing code was developed in collaboration between THORN Lighting Limited and Flow Simulation Limited.

The resulting model of a high pressure discharge lamp in cylindrical geometry has been used to produce 2-D results which closely match our experimental measurements and similar predictions by other authors.[2,3]

The asymmetry of the isotherms due to convective flows is clearly represented and so are the positions of the hot spots in the vicinity of the electrodes and the cool spots on the arc tube walls. The predicted arc centre and wall temperatures are also in agreement with measurements and other predictions. The use of the model to carry out 3-D calculations is discussed elsewhere.[4]

References

[1] L.Cifuentes, Selecting process modelling software (Parts I and II), Chemical Engineering (USA), November 9 and December 7, 1987.

[2] E.Fischer, Modelling of low power discharge lamps, 4th Int. Symp. on the Sc. and Technol. of Light Sources, Karlsruhe, April 1985.

[3] W.Shyy and J.T.Dakin, 3-D natural convection in a high pressure mercury discharge lamp, 5th.Int.Conf.Numerical Methods in Thermal Problems, July 1987, Montreal, Canada.

[4] L.Cifuentes and D.A.J.Mottram, 3-D computer predictions of a high pressure discharge lamp, This Conference, 1989.

COMPUTER PREDICTIONS OF FLOW PATTERNS AND TEMPERATURE CONTOURS IN HIGH PRESSURE DISCHARGE LAMPS

L.Cifuentes

Lamp Research, THORN Lighting Limited, Light Sources Division,

Melton Road, Leicester LE4 7PD, U.K.

The development and improvement of high pressure discharge lamps depend on precise knowledge of the interaction between lamp design and the physical and chemical characteristics of the discharge. Temperature distributions are of great importance both to the efficiency of the light source as well as to its lifespan.

A computer model of fluid flow, heat transfer and electric field in a high pressure discharge lamp [1] has been shown to produce satisfactory predictions. It has now been used in order to predict flow patterns and temperature distributions as a function of the gas pressure, the geometry and material properties of the arc tube, the tube ends and the electrodes.

The input data and the studied variable ranges are fully defined in the paper.

The results show that important lamp properties, such as the position of the cool spot and the temperature distribution at the arc tube ends are sensitive to the studied variables. Comparison with experimental evidence shows that the model predictions are realistic.

An important conclusion is that computer models can be used as powerful time- and money-saving tools for lamp development, by dramatically reducing the range of experimental work.

References

[1] L.Cifuentes, D.A.J.Mottram and S.Vasquez, A computer model of a high pressure discharge lamp based on the FLUENT code, This Conference, 1989.

THREE DIMENSIONAL COMPUTER PREDICTIONS OF A HIGH PRESSURE DISCHARGE LAMP

L.Cifuentes and D.A.J.Mottram

Lamp Research, THORN Lighting Limited, Light Sources Division,

Melton Road, Leicester LE4 7PD, U.K.

The development of a model of a high pressure discharge lamp and some 2-D calculations have been presented elsewhere.[1] The code involves the simultaneous solution of the fluid flow, heat transfer and current continuity equations with appropriate boundary conditions. 2-D calculations, however, cannot tackle the essentially 3-dimensional problem of predicting the flow patterns and temperature contours for a lamp running in positions other than vertical.

It is well known that the position of the plasma changes as the arc tube is tilted and the phenomenon of arc bowing manifests itself, changing the shape of the isothermal surfaces in the plasma, tube walls and ends.

The model allows for 3-D calculations and these were carried out for a cylindrical arc tube. Full details of the input data and the calculation of parameters for the radiation model are given in the paper.

The results agree with experimental observations as well as with those obtained by Shyy and Dakin.[2] The arc bowing in several positions was established, together with flow patterns and temperature contours.

The way to tackle problems arising from the model definition and numerical methods and further work required to improve on current predictions are also discussed.

References

[1] L.Cifuentes, D.A.J.Mottram and S.Vasquez, A computer model of a high pressure discharge lamp based on the FLUENT code, This Conference, 1989.

[2] W.Shyy and J.T.Dakin, 3-D natural convection in a high pressure mercury discharge lamp, 5th Inf.Conf.on Numerical Methods in Thermal Problems, July 1987, Montreal, Canada.

Influence of the two-temperature hypothesis on the electrical characteristics of high pressure mercury discharge plasma.

S.Couris, G.Zissis and J.J.Damelincourt
Laboratoire de Décharges dans les Gaz (URA-277 CNRS)
118, rte de Narbonne, 31062 Toulouse, FRANCE

The high pressure mercury discharge is the most investigated in lamp research (1, 2). We have therefor chosen it to examine the consistency of the two temperature plasma approximation with electrical and spectroscopic measurements.

In the case of two lamps with mercury pressures 2.03 and 4.96 bars, operated at 50Hz the determination of the time evolution of the radial temperature profiles has been performed by using time resolved measurements of the total intensity of 491.6nm and 577nm optically thin lines. Absorption has been taken into account when necessary.

Instantaneous measurements of arc current and voltage have also been carried out through the ac cycle. Using the temperature measurements, electron density has being calculated via Saha equation in the assumption of L.T.E. (electron temperature equal to gas temperature). The arc current is then given by:

$$I = 2\pi E e \int_0^R r \, n_e(r) \, \mu_e(r) \, dr \qquad (1)$$

where E is the electric field strength, e is the electron charge, n_e is the electron density along the radius and μ_e the electron mobility in the mercury. Electric field E is deduced from instantaneous voltage measurements, the experimental value of the electrode potential fall (3) has been taken into account in the electric field calculation. The electron mobility has been calculated by using the kinetic theory formalism (4). In this calculation we have taken into account Hg/e^-, Hg^+/e^- and e^-/e^- interactions, the cross-sections of which these have been taken from the literature (5, 6).

In order to explain the differences between the calculated and experimental values of arc current, we have introduced the concept of a two temperature plasma. Then, by means of the generalized Saha equation (7), the electron density values along the radius, as calculated from simple ionization equilibrium, have been adjusted in order to reproduce the experimental arc current evolution (8).

An effective value of the ratio $\theta = T_e/T_g$ has thus been calculated. We have found that the time modulation of θ is in phase with the electrical field as shown in fig.1. This may be explained by the fact that once the electric field reaches its maximum value and therefore electron accelaration is maximal, the product of electron density by the density of neutrals becomes minimal. The occuring energy transfer between electrons and heavy particles is thus ineffective. This means that an inefficient thermalization takes place in the discharge plasma.

In fig.2 we show the effective values of θ for the two mercury pressures studied. We may roughly conclude and in a rather qualitative manner that thermalization is more efficient for the lower pressure of mercury. We may explain this remark by taking into account, that in this case, much more electrons are able to thermalize the heavy particles than in the case of the higher mercury pressure. These qualitative results are in agreement with multiple parameters model (9) where the excitation temperature of the Hg-6^3P levels is estimated to be greater than under L.T.E., in the case of lower mercury pressures.

A more complete analysis, including temperature and pressure measurements, will soon follow.

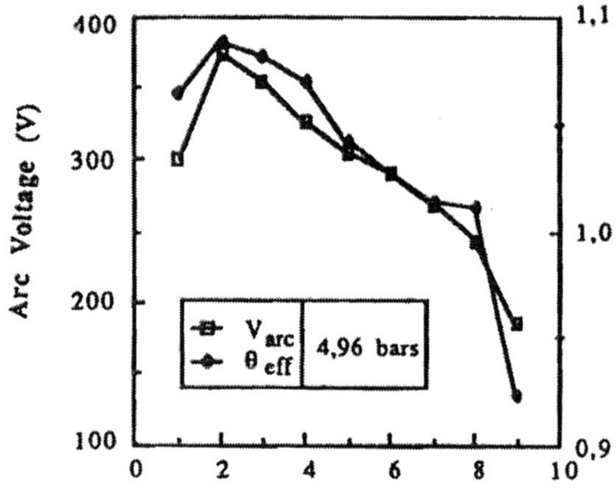

figure 1.

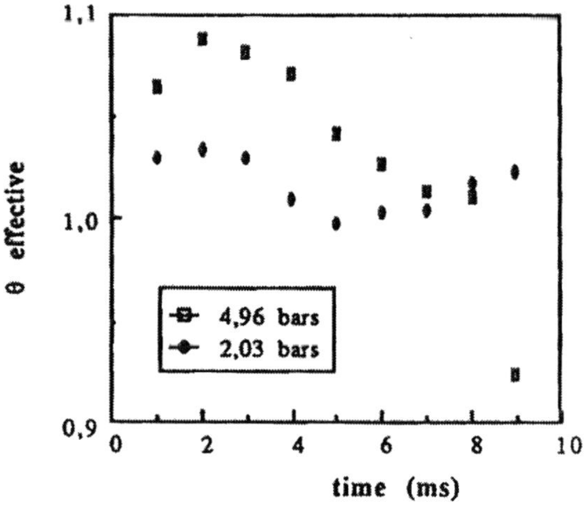

figure 2.

(1) Elenbaas, W., The High Pressure Mercury Vapor Discharge, North Holland Pub.Co., Amsterdam, 1951.

(2) Waymouth, J.F., The Electric Discharge Lamps, MIT Press, Cambridge, 1971.

(3) Soriano, C., Thesis 3ème cycle, Université Paul Sabatier, Toulouse, 1987.

(4) Soriano, C., Aubes, M., et al, Revue Général de l'Electricité, 7, 1-19, 1988.

(5) Rockwood, S.D., Phys. Rev. A, 8, 2348-2358, 1973.

(6) Mitchner, M., Kruger, C.H.Jr, Partially Ionized Gases, John Wiley and Sons, New York, 1973.

(7) Prigogine, I., Bull. Cl. Acad. R. Belg., 26, 53, 1940.

(8) Couris, S., Zissis, G., et al, XIX I.C.P.I.G., Belgrade, 1989 (to be published).

(9) Karabourniotis, D., Couris, S., et al, J. Appl. Phys., 1989 (to be published).

S.Couris is indebted to *European Economic Communities* for partial support.

Determination of the total escape factor in high pressure mercury lamps

J.J.Damelincourt, E.Drakakis and J.Salon
Laboratoire de Décharges dans les Gaz (URA-277 CNRS)
118, rte de Narbonne, 31062 Toulouse, FRANCE
E.Roaux
Cie Philips Eclairage, 4, Rd.Point du Pont de Sevres, 92516 Boulogne Billancourt

In the development of the high pressure lamps one of the main difficulty lies in the estimation of the power repartition in the emitted spectrum, because of the strong radiation trapping in the discharge plasma. We have tied, on the basis of Bartels and Multiple Parameters formalisms, to obtain a simple (but physically significant) evaluation of the total escape factor θ for both strongly and weakly absorbed spectral lines. In this rough evaluation we consider a parabolic temperature profile in the plasma.

We define the total escape factor θ of a spectral line as the ratio of the total radiative flux over the total flux without absorption :

$$\theta = \frac{F_l}{\frac{hc}{\lambda_o} A_{ea} N_{e,l}} \quad (1) \qquad \text{where } F_l = \int_{\Omega=4\pi} I_\phi d\Omega = \int_{\Omega=4\pi} (\Sigma L_T)_\phi \cos\phi d\Omega \quad (2)$$

The geometry of the discharge is given by the fig.1.

Calculation of the total escape factor:

• Bartels formalism (1):

Starting from the spectral radiance L_λ along a line of sight in a direction perpendicular to the axis :

$$L_\lambda = S_\lambda^{max} M Y(\tau_o) \quad (3)$$

Under the assumption of a constant Lorentz profile, one obtains for the total radiance L_T along of the same line of sight

$$L_T = \frac{hc}{4\pi\lambda_o} A_{ea} N_e \frac{\sigma_B(\tau_{max})}{\pi\sqrt{\tau_{max}}} \quad (4)$$

τ_{max} is the optical thickness in the center of the line profile. The value of $\sigma_B(\tau_{max})$ depends slightly on τ_{max} for $\tau_{max}>15$. Details of the calculations is given in reference (2).

• Multiple Parameters formalism:

L_λ is now written as $L_\lambda = S_\lambda^{max} \frac{1}{n} Y(\tau_o)$ where $Y(\tau_o) = \tau_o e^{-\tau_o/2} \int_0^1 \frac{1}{n} \xi^{n-1} \cosh\left[\frac{\tau_o}{2}(\xi-1)\right] d\xi$

and $\xi = 2\int_x^{x_o} \frac{n_a(x)}{N_a} dx$ with $N_{a;e} = \int_{-x_o}^{+x_o} n_{a;e}(x) dx$. If we point out that $n = \frac{2n_e(\xi)}{n_a(\xi)} \frac{N_a(\xi)}{N_e(\xi)}$ we obtain for L_T the same expression as (4) with $\sigma_{PM}(\tau_{max})$ instead of $\sigma_B(\tau_{max})$.

Taking into account expressions (1), (3) and (4) the total escape factor can be written as:

$$\theta = \frac{0.874 \langle g_A \rangle \sigma(\tau_{max})}{\sqrt{\pi\tau_{max;d}}} \quad (5)$$

$\langle g_A \rangle$ is the mean value of a function $g_a(y)$ which describes the integration of L_T over a right section of the lamp, $\tau_{max;d}$ is the optical thickness in the center of the line profile in a diameter of the discharge.

For weakly absorbed line the Multiple parameters and Bartels $Y(\tau_o)$ functions degenerate and $\sigma(\tau_{max})$ can be written

$$\sigma(\tau_{max})=\pi\tau_{max}^{1/2}\left\{1-\frac{\tau_{max}}{4}+\frac{\tau_{max}^{2}}{16}\right\} \qquad (6)$$

The above calculations were applied in the case of the self-reversed visible mercury lines emitted by two high pressure lamps in a 50 Hz ac current. The temperature distribution was determined experimentally using an optimisation procedure for the emission of the Hg-577 nm non-absorbed line. The diametral optical thickness in the line center, has been evaluated by using the distance between the self-reversed line maxima under the assumption of a constant Lorentz profile. Flux measurements were carried out by using an Ulbricht's sphere, calibrated by an incandescent lamp. Table I summarize the experimental (exp) and theoretical (B and PM) results for two operating instants (at current zero crossing, 0ms, and at 5ms after the zero crossing of the arc voltage). Lamp power was 1030W and operating pressure of Hg was 2.7 bar (lamp D3) and 5.4 bar (lamp D5).

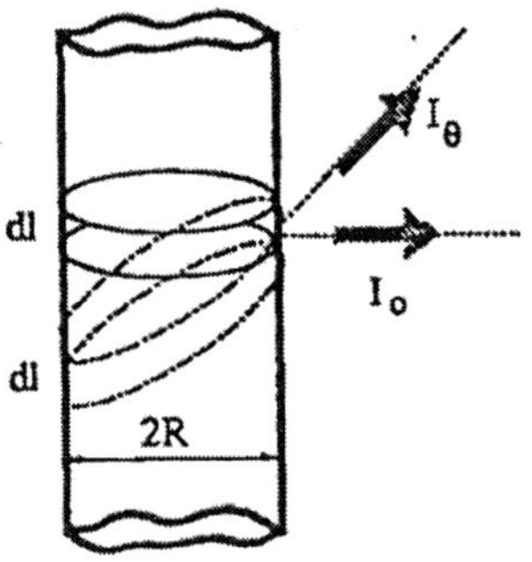

Fig.1. Discharge geometry

Table I

	λ(nm)		546.1	435.8	404.7
Method	t(ms)	lamp			
B	0	D3	0.184	0.185	0.183
PM	0	D3	0.217	0.163	0.210
exp	0	D3	0.220	0.163	0.206
B	5	D3	0.137	0.115	0.129
PM	5	D3	0.119	0.092	0.117
exp	5	D3	0.113	0.095	0.116
B	0	D5	0.194	0.148	0.176
PM	0	D5	0.218	0.171	0.222
exp	0	D5	0.232	0.165	0.210
B	5	D5	0.149	0.119	0.137
PM	5	D5	0.136	0.103	0.130
exp	5	D5	0.134	0.110	0.139

(1) Bartels, H., Uber Linienemission aus Inhomogener Schicht. I. Teil., Zeit. für Physik 125, 597, 1949.

(2) Drakakis, E., Thesis, Université Paul Sabatier, Toulouse, 1988.

A LONG LIFE METAL HALIDE LAMP WITH VARIABLE ELECTRICAL POWER

K.-J. Dietz, A. Schönlein

W.C. Heraeus GmbH, Produktbereich Original Hanau,
D-6450 Hanau

A high pressure mercury metal halide lamp with iron doping for film printing applications which reaches a life more than 6000 hours without intensity loss has been developed. During the life test the lamp has been switched in a fixed cycle from stand by mode to exposure mode (150 W ... 1000 W).

To obtain best conditions for the operation of the lamp particular for the halide cycle and the maximized intensity it is necessary that on each power level the metal halide is essentially vaporized. These determines the geometry of the arc tube, the design of the electrode, the choice of the chemical fillings and the matched working conditions.

We will present the spectral power distributions at different electrical power values as well as data concerning the bulb geometry, electrode design, doping and lifetime curves.

A HG-HIGH PRESSURE LAMP WITH IMPROVED UV-C OUTPUT

K.-J. Dietz, A. Schönlein, W. Schwarz

W.C. Heraeus GmbH, Produktbereich Original Hanau,
D-6450 Hanau

High pressure mercury lamps are widely used in the field of UV-curing and drying of printing inks. They find less application in the field of sterilization and decontamination. But the new improved UV-C light source with high efficiency enables an economic application in this field and further more leads to improvements in the already established applications.

Common high pressure lamps emit the well-known Hg-line spectrum in the wavelength region above 240 nm. These lamps show an efficiency of about 15 to 20 % in the UV-C region (190 ... 280 nm) in respect to the electrical input power.

We present a new developed lamp type with improved UV-C efficiency up to 30 %. This could be achieved by utilizing special excimer processes which are responsible for the UV-continuum.

The spectrum of the common lamp type in comparision to the new developed one together with operational data will be shown.

IONIC DEMIXING IN METAL HALIDE LAMPS

William M. Keeffe,
Richard W. Liebermann,
Harold L. Rothwell, Jr.
GTE Products Corporation
Danvers, MA 01923

In this paper, we present the results of ab initio calculations of the ionic mobilities of Na^+, Sc^+, Hg^+, I^+, and I^- relevant to drift effects in sodium-scandium metal halide lamps, and compare the resulting demixing velocities with measurements[1] on a series of 400W arcs operating at various mercury pressures.

Chapman-Enskog approximations[2] are used to determine the diffusion coefficients and thermochemical equilibrium calculations provide the plasma composition. The mobilities, determined from Blanc's law, are shown in Fig. 1.

Three M400U metal halide lamps were operated horizontally at 400 watts. A polychromator[3] was used to monitor the signals from optically thin lines of Na, Sc, and Hg. Phase detection was used so that the signals could be resolved in time relative to the cathode and anode phases of the electrodes. The three lamps were prepared with normal, 1.5X and 2X mercury dose, respectively.

As the mercury dose is increased, the core temperature also increases and the luminous column becomes more constricted and bowed. The arc length was estimated by accounting for the bowing.

The field strength was determined two ways: (a) The field strength was determined based on the operating voltage, Vrms, and the observed arc length, AL; (b) The field strength was estimated based on the increase in the mercury dose. The axial field strength was assumed to increase as the square root of the increase in the mercury dose.

The results are shown in the following table:

Hg	Vrms	AL	$\langle E_z \rangle_a$	$\langle E_z \rangle_b$	D_{pp}	$\langle v_m \rangle$	u_i	v_c
1.0	130	4.5	26	26	.9+/-.1	108	8	208
1.5X	175	5.0	33	32	.7+/-.1	84	6	198
2.0X	208	5.5	38	36	.4+/-.1	48	4	152

where AL = arc length (cm), E_z = axial field strength by methods a and b, D_{pp} = displacement (cm) between peak cathode and anode cycles, $\langle v_m \rangle$ = measured average drift velocity, u_i = ionic mobility (cm^2/volt.sec) and v_c = calculated drift velocity.

As expected, the measured drift velocities depend inversely on mercury dose, providing evidence that the spectroscopic technique is sensitive to ionic movement. The present measurements provide an average which should be less than the calculated velocities. An improvement between the experiment and calculation is expected when acceleration effects are represented by an equation of motion for the ions. In addition, the measurements reflect some radial motions which were not isolated in this experiment.

References

1. Waymouth, J.F.; Keeffe, et al, 1975. Proc. 1st Symp. Incoherent Light Sources, Loughborough, U.K.

2. J.O. Hirschfelder, C.F. Curtiss, and R.B. Bird, "Molecular Theory of Gases and Liquids", John Wiley & Sons, New York, 1967.

3. Rothwell, Jr., H.L. and Keeffe, W.M. 1981. J of the IES, Vol. 10 (1):40-46.

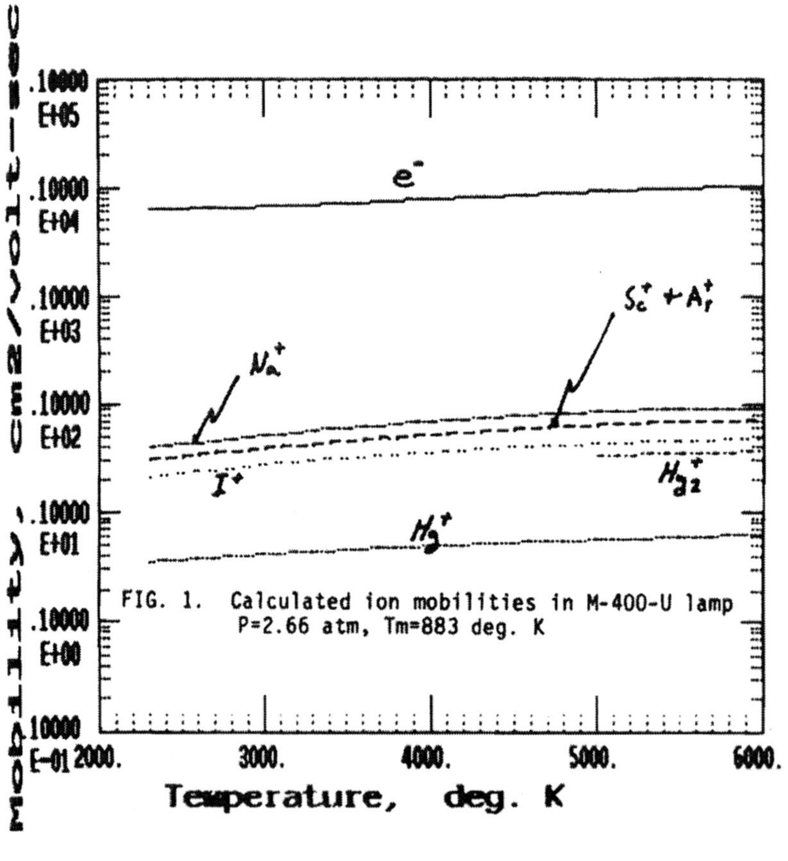

FIG. 1. Calculated ion mobilities in M-400-U lamp P=2.66 atm, Tm=883 deg. K

A SIMPLIFIED RADIAL TEMPERATURE PROFILE MODEL FOR WALL STABILIZED ARC DISCHARGES AND ITS APPLICATION TO THE MODELLING OF RADIATIVE SPECTRA

R. W. Liebermann
GTE Products Corporation
Danvers, MA. 01923

An expression, $T(R) = A_0 + A_2*R^2 + A_3*R^3 + A_4*R^4$, is found suitable for modelling radial temperature profiles of cylindrical shaped arc columns stabilized at a wall radius: $R = r/r_w = 1$. The model is constrained to satisfy: (1) an observed plasma temperature at the central core region, (2) an observed plasma temperature at the arc column's surface, (3) a known thermal heat loss at the wall, and (4) a measured rms I/E ratio. Thermal and electrical conductivity values are required material properties. The model when coupled to radiant heat transfer calculations, eliminates need of iterative solutions to the arc energy balance equation (no axial variation) and thus permits a relatively fast, "first order" calculation of spectral power distributions. Comparisons between simulated and measured power distributions of metal halide lamps help support use of this model.

This paper describes a method for solving the above temperature model and demonstrates its application in modelling and analyzing radiative spectra of metal halide lamps. Sample results are shown in figures 1 through 4. Also included is a description of methods used to calculate the plasma's transport properties, spectral absorptivities and chemical species' concentrations. In addition, several observations regarding the use of "radial demixing" theory in absence of convective constraints is reported.

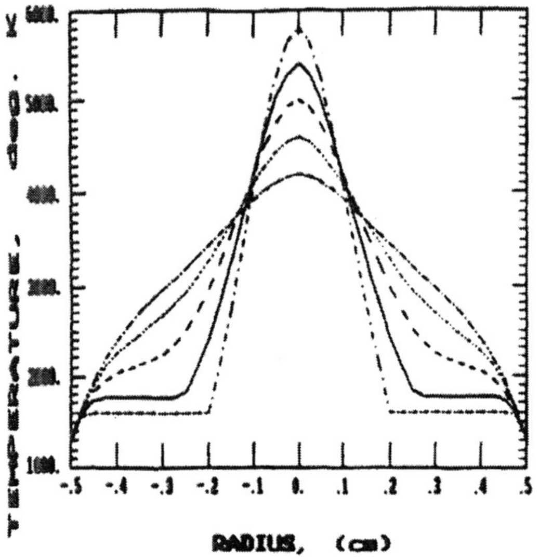

Fig. 1. Variation of Profile Shape vs T(0) at Fixed I/E Ratio

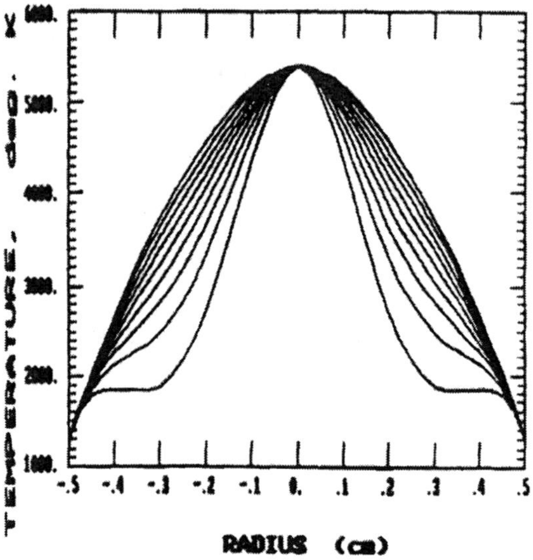

Fig. 2. Variation of Profile Shape vs I/E at Fixed T(0)

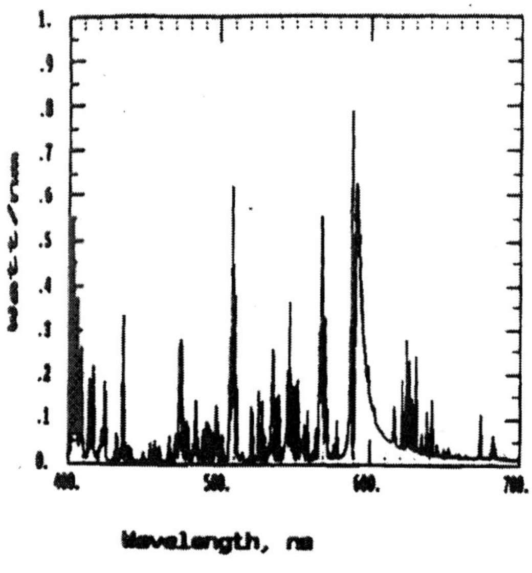

Fig. 3. Simulated Instrument Broaden Spectra

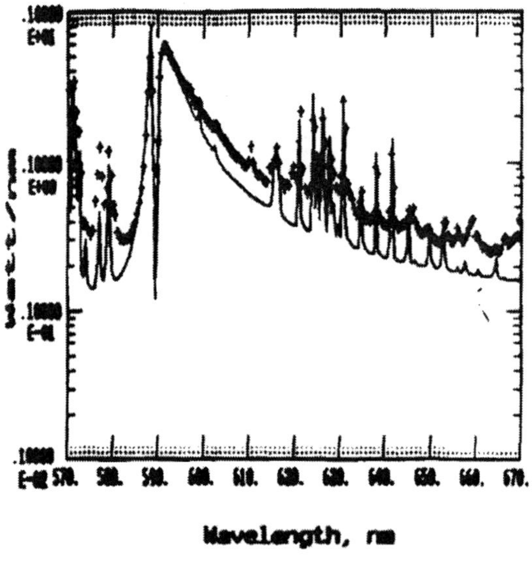

Fig. 4. Simulated Instrument Broaden Spectra and Measured Data

DEVELOPMENT AND APPLICATION OF A 1-D MODEL OF A METAL HALIDE DISCHARGE LAMP INCLUDING DETAILED SPECTRAL PREDICTIONS

D.A.J.Mottram, L.Cifuentes and S.A.Mucklejohn

Lamp Research, THORN Lighting Limited, Light Sources Division,

Melton Road, Leicester LE4 7PD, U.K.

A radial model for a cylindrical metal halide discharge has been developed in our laboratories. It involves the calculation of the chemical composition, the thermal and electrical conductivities of the plasma, the radiation transport and the energy balance in order to determine an accurate radial temperature profile which allows the prediction of a detailed spectrum.

The model is based on previous work carried out for high-pressure sodium lamps [1] with the necessary extensions to allow multiple emitters, radiation continua and line cut-offs.[2]

Experimental work has been carried out to establish the modelling parameters for radiation continua and spectral lines in those cases where no published information was available.

A critical aspect of the model is the correct prediction of spectral features, as omission or unchecked addition of radiative processes make it impossible to obtain an accurate energy balance, despite the fact that the predicted spectrum may closely resemble the experimentally obtained one.

On the other hand, a correct energy balance and an apparently good predicted spectrum may be insufficient to calculate the right light output characteristics. Only precisely calculated lineshapes and continua can give a realistic colour rendering index and colour temperature.

The necessary data input, calculation procedure, predicted spectra and a detailed discussion of the results are given in the paper.

References

[1] P.L.Denbigh, B.F.Jones and D.A.J.Mottram, Relationship between efficacy, arc tube temperature and power dissipation in a high pressure sodium lamp, J.Phys. D: Appl. Phys., 16, 2167-2180 (1983).

[2] J.T.Dakin, Energy balance of the high-pressure mercury discharge with sodium- and scandium-iodide additives, J. Appl. Phys., 62, 79-87 (1987).

A New Metal Halide Lamp for Color LCD Projectors

M. Narita, Lamp research & development

USHIO INC. HIMEJI-SHI JAPAN

For the light source of the color LCD projector, a high efficacy, small variation in the color temperature and a good color rendition are required. Metal halide lamps are the most suitable. The short arc lamps filled with rare earth metal halides have good characteristics. It is very difficult to design the metal halide lamp because the envelope is gradually subject to devitrification and the color temperature decreases during operation. In this situation, we evaluated the characteristics of the lamps filled with some types of rare earth metal halide and found Lutetium (Lu) iodide had the best performance against devitrification and a decrease in color temperature. In order to enhance the red part of radiation, Lithium (Li) iodide was added. Li was found to be effective for enhancement of the red light without devitrification, if the ratio of Li to Lu and an amount of iodine are controlled.

The lamp filled with Lu-Li iodide maintains good characteristics even after 1000 hours of operation. These good characteristics may be interpreted in terms of the formation of a stable complex molecule in the lamp. It is possible to modify the spectral distribution by adding indium and/or thallium halide, which emits strong blue and green radiation respectively. Some characteristics of the lamp are shown in the following:

Lamp wattage	:	400w	Life time	:	> 1000hrs
Lamp efficacy	:	80lm/w	Luminous flux (*1)	:	125lm
Color temperature	:	5000k	Chromacity (*1)	:	x=.358
Arc length	:	7mm			y=.371

*1; through the white LCD, on the screen (49*37 inch)

NUMERICAL SIMULATION OF RARE GAS HIGH PRESSURE SHORT ARC LAMPS

A.C. Palacin and M. Neiger

Lichttechnisches Institut, Univ. of Karlsruhe, D-7500 Karlsruhe, FRG

Rare gas short arc lamps are extremely high radiance spot light sources used e.g. for projection purposes, fiber optic illumination systems and possibly incoherent pumping of cw dye lasers. The high brightness plasma spot of these lamps is located very close to the cathode tip and strongly constricted by a high velocity gas flow generated by the interaction of the plasma current with its own magnetic field. An effort to predict plasma characteristics, spectral radiance and scaling laws with respect to operating pressure, plasma current, electrode distance and shape has been undertaken by a numerical simulation of such lamps.
Modelling of the stationary plasma state for LTE conditions requires the simultaneous solution of the energy balance equation (including convective, thermodiffusive and radiative terms), the momentum balance equation (including nonlinear convective, viscous and self magnetic forces), Ohm's law and the conservation equations for mass and charge. Boundary conditions for the lamp plasma within a closed quartz envelope are wall and electrode temperature distributions together with the fill pressure and the operating current.

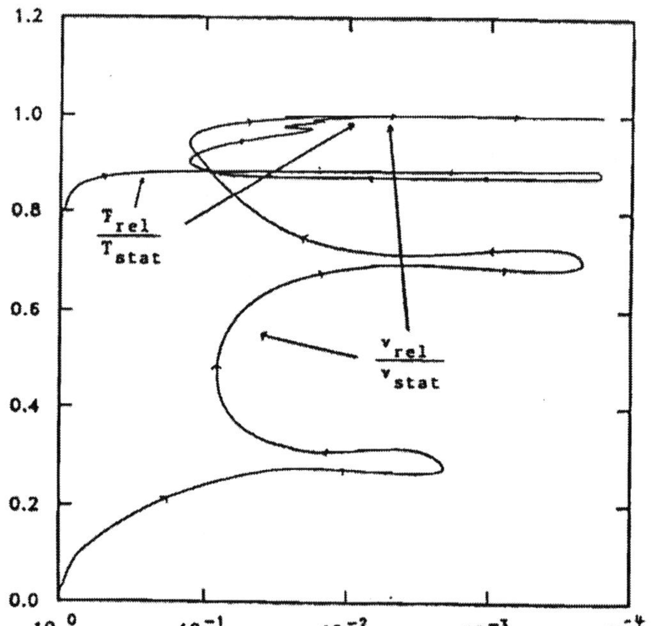

Fig. 1: Ratios of momentary temperatures and velocities compared to final ones during iteration at grid points of maximum temperature and axial velocity. The horizontal axis corresponds to rel. deviations from zero of energy and momentum balance respectively.

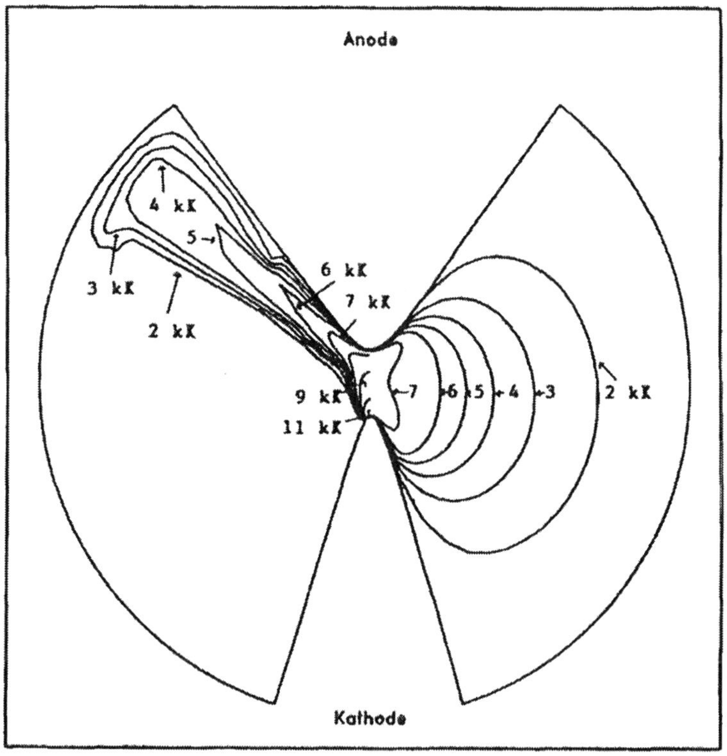

Fig. 2: Temperature field within lamp envelope with (left) and without gas-flow (right).

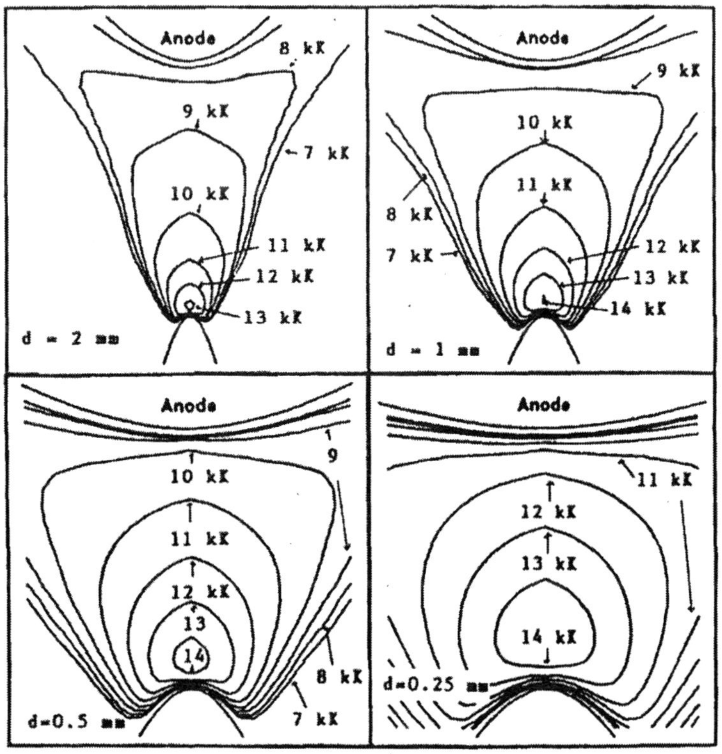

Fig. 3: Temperature fields between cathode and anode for different electrode distances (Xe, p=8 MPa, I=15 A, wall located 5 mm from axis).

THE FUTURE OF METAL-HALIDE LAMPS FOR PHOTOCHEMICAL PROCESSES

G.S.Sarytchev, L.B.Prikupets, G.N.Gavrilkina and
L.A.Smorodinova
All-Union Research Lighting Institute (VNISI)
Prospect Mira, 106, Moscow, 129301, USSR

Irradiating technology embraces a large class of photochemical and photobiological optical radiation receptors (excluding the eye). The existence of a variety of receptors calls for matching requirements to be met by irradiating systems and, in particular, radiators. Along with the traditional light sources, there is a need for creating new actinic lamps having a special radiation composition and an enhanced reliability.

Radiation in a number of optical spectral regions can be produced with the use of emitting additives, which presents problems in the course of lamp manufacture. Among such additives one can mention alkali metals (chemically active with respect to the fused silica lamp envelope glass, oxygen and atmospheric moisture), elements with a low partial of halide vapours, etc. Based on the study of the physico-chemical processes in the discharge and on the lamp walls, a variety of promising approaches to the lamp filling compositions have been considered to enlarge the range of possible lamp types. For this purpose use can be made of alloys containing alkali metals as well as complex metal halide compounds.

The lamp can be filled with an intermetallic compound, e.g., Li_2Sn_3 or Ce_4Cs which then gives rise to a halide directly in the lamp. This procedure can be effected during standard manufacturing processes because such a compound exhibits lower chemical activity. Moreover, the addition of an intermetallic compound allows to control the synthesis of various substances in the discharge in the near-wall low-temperature layer and hence the additive radiation yield.

Using complex halide compounds permits to control the saturated vapour pressure in the system at a lower halide activity and dispense a few additives at the same time. An increased saturated vapour pressure of the complex compounds gives rise to a higher concentration of the emitting additives having a relatively low vapour pressure. This concentration governs the discharge efficiency in the actinic region of the spectrum, e.g., in the $LiI-AlI_3$ or $CsI-AlI_3$ system. Formation of a complex compound whose activity is lower than that of some halides such as alkali metal halides in the near-wall region during ring closure decelerates the process of quartz degradation. Solid phase complex compounds, e.g., $CsHgI_3$, are less active with respect to oxygen and atmospheric moisture therefore they can be dispensed in open air. A preliminary analysis of the compounds allows to dispense a minimum amount of such additives to a high accuracy.

Data are available concerned with the results obtained with the use of ultraviolet lamps for photocopying and plant growth applications, solar light simulators, lamps for TV and movie lighting, etc.

RUN-UP CHARACTERISTICS AND PHOTOMETRIC ASPECTS OF A METAL-HALIDE LAMP FOR A MOTORCAR HEADLIGHT SYSTEM

D. van Pijkeren

Philips Lighting B.V., Eindhoven, The Netherlands

In recent years a lot of work has been done on the development of a low-wattage metal-halide lamp for automotive applications [1,2,3]. The most important subject covered in these studies is the run-up behaviour of the gas-discharge lamp.

If no special precautions are taken the run-up time, which is defined as the time between ignition of the lamp and the moment that 30% of the nominal luminous flux is reached,is as long as one minute for a 35 Watt metal-halide lamp. The reason is that in the run-up period, in which the bulb is still cold, only a fraction of the nominal power is dissipated in the lamp owing to the low lamp voltage. The run-up time can therefore be reduced by increasing the lamp current and the lamp voltage in the run-up period.

A higher lamp voltage of the cold lamp can be realized by applying a high-pressure Xenon filling. In this way the minimum lamp voltage can be increased from approximately 12 volts to more than 20 volts. The use of Xenon has the additional advantage that the Xenon plasma radiates a portion of its energy in the visible region. This phenomenon makes it possible to achieve an instantaneous light output. The amount of Xenon radiation during the run-up period is strongly dependent on both the Xenon pressure and the run-up current. When both a high pressure Xenon filling and an increased run-up current are applied to the lamp, the 50% level of the luminous flux can be reached within 0.5 sec. The run-up characteristic of this type of gas discharge lamp thus strongly resembles that of the incandescent lamp.

The application of high pressure Xenon in the discharge vessel has some negative consequences as well. For instance, the ignition voltage will be strongly increased. Another effect is the enhancement of the convective flow inside the lamp owing to the higher total pressure. As a consequence the lateral displacement of the arc due to buoyance forces will be stronger. This leads to a curved arc which is, from an optical point of view, less attractive. Moreover, a strong bending of the arc can locally increase the quartz temperature and thus reduce lamp life.

The strength of the convective flow is indicated by the Grashof number. Apart from some physical and material constants it contains the total pressure and the inner diameter of the lamp. This inner diameter is thus a parameter to control the arc bending of the arc.

References

1. A. Inouye et. al. Instant Start Miniature Metal Halide Lamp, fourth International Symposium on the Science and Technology of Light Sources ,1986
2. B. de Vrijer, Patent No: 4594529
3. J.M.Davenport and R.L.Hansler, Instant Light for Metal Halide Lamps ,IES Annual Conference 1988.

IMPROVEMENT OF ULTRAVIOLET-RELATED PROPERTIES OF MHN-T LAMPS.

J.A.J.M. van Vliet and W. Verbeek
Philips Lighting B.V., Eindhoven, The Netherlands

The interest in high pressure gas discharge lamps for accent and general lighting in merchandising applications is on the increase. Besides the introduction of a series of metal halide lamps for closed luminaires a few years ago a new version of the high pressure sodium lamp was introduced: the so-called white sodium lamp with tubular envelope (SDW-T). A special feature of this single-ended lamp is that it can be used in open luminaires.

In application where higher colour temperatures are required, the SDW-T lamp family has to be extended with metal halide lamps. The first offspring in this family of metal-halide lamps, which are optically interchangeable with the SDW-type lamps, is the recently developed single-ended MHN-T 70W lamp. Because of its small discharge tube and single-ended compact lamp construction in luminaires intensive light beams of more than 70 kCd can be realized.

Apart from the obvious requirement of explosion safety, the MHN-T lamps must also meet requirements with respect to their ultraviolet related properties if they are to be used safely in open luminaires. Excess ultraviolet radiation can lead to damage to human skin and eyes and can cause the colours of objects to fade.

The directives of the National Institute for Occupational Safety and Health and the National Bureau of Standards may be followed in the design of metal halide lamps for open luminaires.

The maximum time human beings may be exposed to the radiation of a light source which also radiates a small amount of ultraviolet radiation is expressed by the Permissable Exposure Time. This PET-value indicates the maximum time a healthy adult may daily be exposed to a light source giving an illumination level of one kilolux.
The present demand on light sources is that the PET-value should exceed a value of 16 hrs.

A measure for the relative fading of objects exposed to radiation sources is the damage factor. It is defined by the ratio of the spectral distribution of the light source weighted according to the relative spectral damage function and the eye sensivity function respectively. The damage factor of metal halide lamps should be comparable to that of other practical light sources for display and general lighting.

In this contribution we want to discuss the means by which the ultraviolet-related properties of MHN-T lamps can be improved in such a way that the above requirements can be fully met.

NUMERICAL TESTS OF A SIMPLE EMPIRICAL MODEL FOR RADIATION TRANSPORT IN OPTICALLY THICK PLASMAS

P. A. Vicharelli

GTE Laboratories Incorporated
Waltham, MA 02254 USA

and

D. Karabourniotis

Department of Physics
University of Crete, Iraklion, Crete, Grece

One of the earliest studies of optically thick radiation is that of Bartels,[1] who presented an approximate analytic description for the emitted intensity. Bartels' model, however, is only valid at low and intermediate values of the optical depth; this precludes its application in the neighborhood of a self-reversed minimum.[2,3]

Other simple treatments of the optically thick plasma that include a description of the self-reversed minimum make assumptions about the functional form of the internal emission/absorption relationship, i.e., about the source function. One approach (CDK), which evolved from the Cowan-Dieke[4] model for self-reversed lines, has been used in the determination of the plasma temperature,[5] the partial pressure of the emitting species,[6] and more recently, in the study of deviations from local thermodynamic equilibrium.[7-9] However, some of the basic assumptions of the underlying model have not been rigorously tested.

The two main assumptions of the CDK model are (a) a particular functional form for the relative distribution of emitters to absorbers (the excitation function E) is assumed, and (b) the collisional broadening is taken to be independent of position. The purpose of the present work is to report the results of a systematic quantitative study of the limits of applicability of assumption (a) of the CDK model. We use a mercury arc lamp as a test case. A mercury arc provides an ideal example of a system with a monotonic radial distribution of emitting and absorbing atoms, with most of the radiation coming from the arc core.

In this test we define a series of radial temperature distributions given by the formula $T(r) = (T_0 - T_w)(1-[r/R]^2)/(1+S[r/R]^3)+T_w$ where T_0 and T_w are the core and wall temperatures, respectively, r is the radial coordinate, R is the arc tube radius, and S is a shape parameter. As shown in Fig. 1 this parameter may have values 0, 10, and -0.9 to give profiles with parabolic, constricted, and flattened shapes, respectively. Next, we solve the radiation transport equation numerically to calculate the spectral lineshapes for several Hg lines and thus simulate an experiment. The results of this 'experiment' may then be processed according the technique derived from the simple model, which assumes that the excitation function is given by $E(y) = y^n$ where $y = 1 - |\xi|$, ξ is the relative optical thickness, and n is an adjustable parameter. The first test that we consider is that of comparing the absolute intensities of the numerical and the model lineshapes. In other words, we investigate how well the empirical relationship given above fits the exact excitation function. We have found that for parameters typical for an Hg arc ($T_0 = 6\times10^3$ and $T_w = 10^3$ K) the intensities calculated according to the CDK model may be off by as much as 50%, relative to the exact results, at line center, where the optical thickness is a maximum. In contrast, at the peak of a self-reversed maximum we find the the model is accurate to a few percent.

A different, and perhaps more relevant, kind of test employs relative lineshape data. When intensity ratios are used, the systematic deviations exhibited by the simple model cancel out to a large extent. Consider one of the quantities that can be extracted from relative intensities using the CDK treatment–the radially integrated density of absorbers N_a. Since the radial temperature distribution is known, this quantity may be easily calculated. Then we can extract a value $N_a(model)$ from the simulated experimental data according to the CDK method and compare the two. The results of such a comparison are summarized in Table I in terms of the ratio $\rho = N_a(model)/N_a(exact)$. For a parabolic temperature profile, the simple model gives N_a values that fall within 10% of the exact ones. This case is closest to Hg arcs, whose temperature profiles resemble parabolas. The shapes in the other two cases are somewhat extreme and could be regarded as upper and lower limits of all possible Hg T(r) profiles. Consequently, the observed deviations, of up to 28%, shown for these two cases are likely to represent the maximum uncertainty introduced into the analysis by the model itself.

REFERENCES

[1]H. Bartels, Z. Phys. **127**, 243 (1950); *ibid.* **128**, 546 (1950).
[2]P. A. Vicharelli and L.C. Pitchford, 4th Int. Symp. Light Sources, Karlsruhe, 1986.
[3]P. A. Vicharelli and P. A. Reiser, Phys. Lett. A **125**, 326 (1987).
[4]R. D. Cowan and G. H. Dieke, Rev. Mod. Phys. **20**, 418 (1948).
[5]D. Karabourniotis, J. Phys. D **16**, 1267 (1983).
[6]D. Karabourniotis, S. Couris, J. J. Damelincourt, and M. Aubes, IEEE Trans. Plasma Sc. **PS-14**, 325 (1986).
[7]D. Karabourniotis, Opt. Commun. **61**, 38 (1987).
[8]D. Karabourniotis and S. Couris, Opt. Commun. **67**, 214 (1988).
[9]D. Karabourniotis *et al.*, J. Appl. Phys., in press.

TABLE I

Accuracy of the CDK model for several types of radial temperature distributions.

Profile	$\rho(6^3P_0)$	$\rho(6^3P_1)$	$\rho(6^3P_2)$
Constricted	1.13	0.94	1.04
Parabolic	1.05	0.92	0.98
Flattened	0.82	0.72	0.77

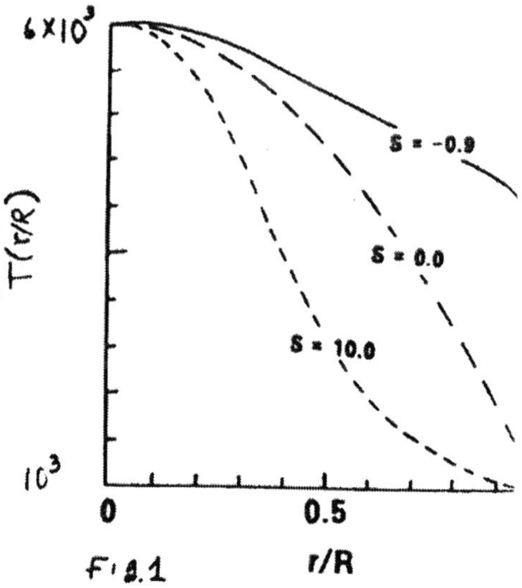

Fig.1

COMPUTATIONAL FLUID DYNAMICS OF HIGH INTENSITY ARCS

P. A. Vicharelli and T. Fohl

GTE Laboratories Incorporated

Waltham, MA 02254

We have carried out computational experiments on the fluid dynamics of high-intensity-discharge arc lamps. The discharge plasma is viewed as a viscous, compressible, laminar fluid in local thermodynamic equilibrium. The calculations are based on the numerical solution of the energy,

$$\rho C_p \vec{V} \cdot \vec{\nabla} T - \vec{\nabla} \cdot (\kappa \vec{\nabla} T) = \sigma E^2 - U_R$$

momentum,

$$\rho(\vec{V} \cdot \vec{\nabla})\vec{V} - \vec{\nabla} \cdot (\eta \vec{\nabla} \vec{V}) = -\vec{\nabla} P - \rho g \hat{e}$$

and mass

$$\vec{\nabla} \cdot \rho \vec{V} = 0$$

conservation equations for the plasma fluid, coupled with the radiation transport, current continuity, and chemical equilibrium equations. In addition, heat transfer from the plasma to the enclosure has been included in the model by means of a fluid-solid energy balance. Here ρ is the plasma density, C_p is the specific heat, $\vec{V}$ is the convective velocity, κ is the thermal conductivity, T is the temperature, σ is the electrical conductivity, E is the magnitude of the electric field, U_R is the net emission coefficient, η is the viscosity, P is the pressure, g is the acceleration of gravity, and $\hat{e}$ is a downward unit vector.

Two different approaches are explored in this paper. First, we employ a finite-element formulation of these equations to investigate the behavior of convective circulation patterns and their effect on the plasma and discharge-tube temperature distributions. The finite element method is very convenient in that it allows the treatment of arbitrary-discharge tube geometries. In our implementation, a separate finite-element code is written for each of these equations. The mass conservation constraint is transformed into a pressure equation,[1] and the resulting coupled set of equations is solved iteratively until a self-consistent solution is obtained. Results for Hg arcs simulated at several pressures, power loadings, and arc tube shapes will be presented.

One of the main reasons for choosing an Hg arc for our calculations is that most of the input data, i.e., spectroscopic, thermodynamic, and transport properties, are available in the literature. Therefore this type of arc is a convenient vehicle for testing and validating a fluid dynamics code. However, if we investigate arc lamps with more complex and interesting chemical fills we find that reliable input data is frequently not available. We address this issue in our second approach, where instead of solving the energy balance equation to obtain a calculated temperature distribution, we employ tomographically determined experimental temperatures, and proceed to solve the rest of the fluid flow equations. Tomographic diagnostics are discussed in a companion paper.[2] A great advantage of this dual experimental-computational approach is that the most demanding part of the calculation, the solution of the radiation tranport equation, is eliminated, as is the need to find spectroscopic properties for the chemical constituents of the lamp. Thus, in a sense, flow calculations done this way allow us to visualize the actual fluid flow motion taking place in the arc, as well as the spatial distribution of demixed minority chemical species. Results of this kind of analysis for metal halide lamps will be presented.

REFERENCES

[1]R. J. Schnipke and J. G. Rice, Int. J. Numer. Methods Engrg. **24**, 117 (1987).

[2]T. Fohl and P. A. Vicharelli, *Computerized Tomography of High Intensity Arcs,* preceeding paper.

Predicting high-pressure discharge performance

James T. Dakin
Advanced Technology Department
GE Lighting, Nela Park, Cleveland, OH 44112

March 20, 1989

INTRODUCTION

There is a wealth of devices which fall into the category of high pressure discharge light sources. In terms of power ratings, these range over two orders of magnitude, from about 30 W to about 3000 W. The light comes from atomic lines in some cases, and in others from molecular transitions or recombination. Most of these devices involve quartz arctubes, but some which employ sodium or cesium do not. Most of these devices use electrodes to couple electrical energy to the plasma, but some inductively coupled plasma light sources do not. These various devices are optimized for a wide range of applications having different requirements for color, efficiency, life, and cost. The most common devices are the high pressure mercury, high pressure sodium, high pressure xenon, and metal halide lamps. There are also less well known devices, such as a variety of high pressure discharge light sources used in spectroscopy.

A common set of physics and chemistry principles underlies all of these devices. The fundamental assumption, or approximation, is that of local thermodynamic equilibrium, LTE. With LTE, one can predict the chemistry if one knows the temperature, predict the radiation if one knows the chemistry and temperature, or predict the temperature if one knows the chemistry and radiation. By simultaneously solving these circular relationships one can, in principle, predict the light output and electrical behavior of a specific lamp based on a prescription of its construction and operating conditions. One can also predict many aspects of the condensed phase temperatures and chemistry which determine lamp life. A predictive model of this type can be a powerful tool for evaluating design sensitivities. The relationships can be readily written in a sufficiently general form that they can apply to *any* high pressure discharge light source. The challenge is to sort through the details, and make this science applicable to product development.

HIGH PRESSURE DISCHARGE PHENOMENA

This section will briefly review literature addressing key aspects of high pressure discharge science. The emphasis is on more recent work, work which builds a framework for predictive modeling, and work on the generally more challenging metal halide system. The book of de Groot and van Vliet deserves special attention as a thorough and up to date review of the physics and chemistry of the high pressure sodium lamp, which is probably the best understood of the lamps in question.[1] The earlier books of Waymouth and Elenbaas provide comprehensive but more dated treatments of other high pressure discharge light sources.[2,3]

Fluid phenomena

The high pressure discharge may be viewed as a fluid which obeys principles of energy and momentum balance. Jack and Koedam have provided comprehensive energy balances for a number of high pressure discharges.[4] Experimental diagnostics and model considerations are used to trace the electrical energy entering the arctube through to the energy leaving the arctube as discharge radiation in various bands, discharge thermal load on the arctube wall, and electrode losses. Energy balance data of this type remains an important benchmark for testing models. Recent data from a number of sources provides sensitivity of the high pressure sodium energy balance to variations in various arctube parameters.[5,6,7] The work of Dakin and Gilliard demonstrates a first principles model of the discharge energy balance in metal halide systems with suppressed convection.[8,9] Time-dependent solutions of the energy balance equations have been reported for a number of high pressure lighting discharges operated with pulsed or sinusoidal excitation.[10,11,12,13] These studies show, in one way or another, effects associated with discharge recovery after having been turned off for some period of time.

The discharge fluid becomes more interesting when natural convection brings the momentum balance equations into the problem. Kenty's work on convection in the high pressure mercury discharge remains important as it provides a basic model for the convection, and quantitative measurements of convection velocities.[14] Lowke has provided a numerical approach to solving the coupled fluid equations for vertical arcs.[15] In studies involving numerical models and representative mercury discharges, Zollweg showed that convection plays a key role in arc constriction near the lower electrode, but a much less significant role in energy transport to the wall than the role played by resonance radiation emission and absorption.[16] Using similar numerical techniques, Shyy and Dakin have calculated three dimensional convection patterns in non-vertical mercury arctubes.[17]

Chemistry phenomena

Discharge chemistry brings a whole new dimension to the modeling problem for metal halide systems. For simpler systems, such as high pressure xenon, sodium or mercury, the chemistry generally reduces to solving the Saha equation so as to predict the ionization as a function of temperature for the gas mixture in question. In the metal halide discharge, the situation is much more complicated. Gas properties, such as the electrical conductivity, the radiation emission and absorption, and at high temperatures the thermal conductivity, are dominated by the metal halide additives, and are not a function of temperature only. A key series of papers of Fischer, Stormberg, and Zollweg et al. shows how competition between additive diffusion and convection leads to "de-mixing" or "segregation" phenomena through which the additive pressures can vary enormously from place to place in the discharge.[18,19,20] Diagnostic measurements, such as those of Work or of Allen et al., clearly show the depletion of Na from the upper regions of the vertical discharge.[21,22] The laser absorption measurements of Allen et al. are particularly significant because they trace the sodium groundstate density all of the way from the discharge center, where it is reduced by de-mixing, ionization and the $1/T$ effect, through a broad maximum, to the wall, where it is reduced by iodide recombination. It is comforting that the trends in these data are consistent with the assumptions and predictions of the de-mixing models.

Modeling of condensed phase chemistry cannot be completely separated from discharge modeling. In high pressure sodium and metal halide discharges, vapor pressures are determined by the temperatures, compositions, and locations of condensed phase arctube ingredients. These temperatures, compositions and locations are often influenced in one way or another by the discharge. In addition, condensed phase chemistry involving the electrode and wall materials often determines product life. The discharge often plays a role in transporting the ingredients which participate in this chemistry. An excellent review of chemistry issues in metal halide systems has been provided by Work.[23] As condensed phase chemistry issues are beyond the scope of the present paper, no further attention will be devoted to them.

Radiation phenomena

Radiation is central to the modeling of all high pressure discharge systems. The quantification of radiation is important both because radiation is the "product" which the discharge manufactures, and because radiation plays a dominant role in the energy balance discussed earlier. The challenge comes from the fact that for most systems, much of the radiation comes from reversal regions of self-absorbed spectral lines. This often means that radiation emitted near the discharge center and absorbed in the wings of the discharge plays a key role in determining the discharge profile. Neither optically thick nor optically thin approximations are particularly helpful. Lowke, who developed a numerical scheme for simultaneously solving the radiation transport equations and the discharge energy balance equation, showed that emission and absorption of sodium resonance radiation is particularly important in "fattening" that discharge.[24] Zollweg, as mentioned earlier, made similar observations regarding the high pressure mercury discharge.[16] Dakin and Gilliard showed that resonance radiation plays a less significant role in the metal halide discharge.[9] While the Lowke numerical scheme is formally correct, it consumes a lot of computation time. Various more efficient schemes have been reported by others.[25,26]

Important research has sorted out some the many individual radiation mechanisms which contribute to the total spectrum. The work of de Groot et al. has led to an understanding of the $Na-Na$, $Na-Hg$ and $Na-Xe$ molecular effects which contribute to the wings of the sodium resonance line in the high pressure sodium lamp.[27] Zollweg et al. have described a continuum bremsstrahlung phenomenon which contributes to many high pressure mercury based systems.[28] Stormberg has contributed a scheme for parameterizing pressure broadened spectral lineshapes in high pressure mercury based systems.[29] The recent doctoral thesis of Palacin includes a model of radiation phenomena in the high pressure xenon discharge.[30]

Electrode phenomena

Effects associated with electrodes have remained the greatest uncertainties in modeling the high pressure discharges. Several factors contribute to this. Unlike positive column phenomena, electrode phenomena are fundamentally two-phase and two-dimensional in character, and occur over small distance scales which are difficult to resolve experimentally. Fischer has pointed out that the electrode regions play a proportionately larger role in smaller wattage discharges, and has shown a modeling approach which naturally integrates the electrode phenomena with the positive column phenomena for a high pressure mercury system.[31] This approach involves a coordinate system which provides more resolution in the vicinity of the electrode tips, and a treatment of non-equilibrium phenomena at the electrode-plasma interface. Tielemans and Oostvogels have measured electrode temperatures under a variety of conditions, and related these to an electrode energy balance model.[32] Waymouth has provided a basic model of cathode spot physics which is able to predict why some cathodes operate in a spot mode and others in a diffuse mode.[33] Waymouth has also presented experimental data and a model for the glow-to-arc transition which occurs on lamp starting.[34] Schafer and Stormberg have studied the positive column contraction regions in high pressure mercury discharges, and shown that these regions cool more rapidly than the rest of the positive column when power is removed.[35] Thus these regions make a disproportionate contribution to the discharge reignition voltage.

Local thermodynamic equilibrium

As mentioned earlier, the assumption of LTE is a cornerstone of most scientific investigations of high pressure lighting discharges. In emission spectroscopy diagnostics, LTE enables measurements of temperatures and species densities via well known techniques. In the modeling of radiation and chemistry phenomena, LTE allows one to proceed without knowing a myriad of rate coefficients. Karabourniotis et al. have developed emission spectroscopy techniques for measuring excited state densities without assuming LTE, and have shown evidence for departures from LTE involving the low lying mercury excited states in a high pressure mercury discharge.[36,37] The qualitative arguments in support of these non-equilibrium effects seem plausible, however, there is as yet no quantitative way of including the effects in a discharge model.

NEW RESULTS FROM PREDICTIVE MODELS

Pulsed high pressure sodium spectrum

The high pressure sodium model of Dakin and Rautenberg solves the time dependent energy equation of the one dimensional positive column in conjunction with the radiation transport equations so as to predict the spectrum under various conditions of pulsed excitation.[13] This model has been extended by including the treatment of molecular radiation phenomena developed by de Groot et al.[27] Calculated lamp spectrum are compared with measured lamp spectra in Fig. 1 for an arctube operated on 1 kHz square waves with 100% and 25% duty cycle. With this model one can study the impact on the spectrum of changes in variables such as the pulsing conditions, or the sodium pressure.

Vertical metal halide chemistry

A model of a vertical metal halide discharge has been developed which includes two dimensional effects associated with convection, chemical de-mixing, and the electric field. This model is able to predict temperatures, velocities, chemistry and radiation throughout the discharge from first principles without prior knowledge of any of these quantities. Some results from an earlier version of the model without experimental verification have been published.[38] More recently experimental verification has been provided for a 400 Watt $Hg-NaI-ScI_3$ system, and some of the values in the physics and chemistry data bases have been refined. Plots of the predicted and observed temperatures and Na atomic densities along the discharge center axis appear in Fig. 2. The discharge contraction regions near the electrode tips are found to be accentuated in the metal halide system by the radial de-mixing effect. Tad Rautenberg has played a key role in the spectroscopic measurements associated with this work.

Convection in high pressure mercury discharges

The three dimensional model of discharge convection described by Shyy and Dakin has been extended to the study of non-cylindrical arctubes.[17] Computation time in this model is reduced by assuming that the discharge fluid properties and the radiation term in the energy equation are simply functions of temperature, eliminating the need for iterative chemistry and radiation transfer calculations. The temperature contours and convection pattern for a horizontal high pressure mercury discharge, whose arctube has been bent to conform to the discharge "arc" shape, are shown in Fig. 3. Peggy Chang has been instrumental in developing this three dimensional model.

FUTURE OPPORTUNITIES

The ability to do predictive discharge modeling has been established at GE Lighting and elsewhere. There is every reason to believe that the models will improve and be validated over a broader range of the product design space. They will have a growing impact on product development as they explore new concepts and quantify design tradeoffs. They will operate more quickly and with more confidence.

This increased model usage will generate three needs. The first need is for increased computer power, especially for two and three dimensional calculations. It is likely that this need will be met by predictable trends in the computer industry, which are pretty much outside of the control of those of us doing discharge modeling.

The second need is for more data in basic physics and chemistry areas such as the thermochemistry of discharge species, cross sections associated with the de-mixing problem, and broadening constants for spectral lines. There is a lot that we can and should do to communicate these needs to people who are in positions to fund or carry out basic investigations.

Third, we need to extend our discharge models and diagnostics in areas where current knowledge of basic mechanisms is weak. In the case of LTE departures, for

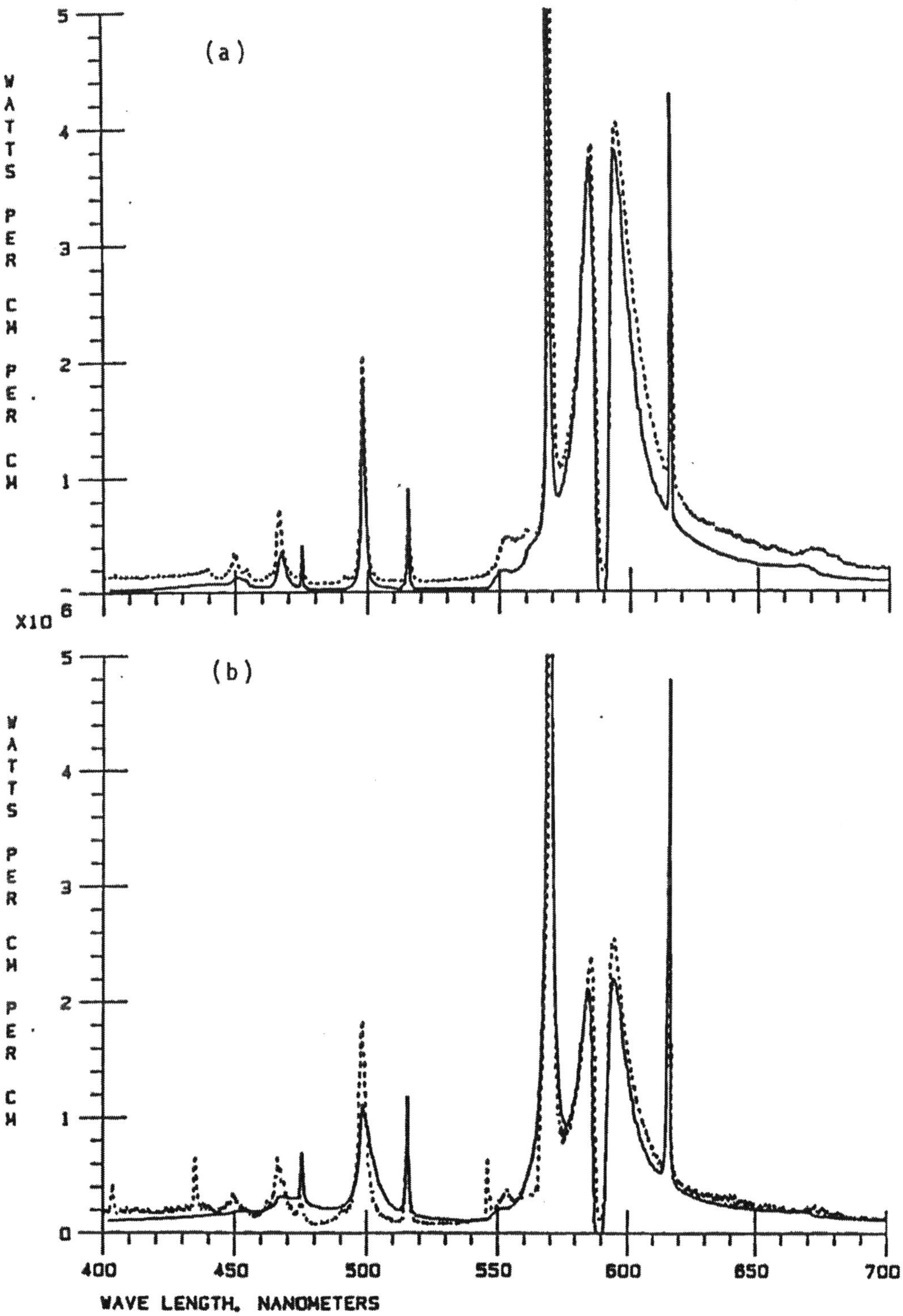

Figure 1: Predicted (solid) and observed (dashed) spectra for a high pressure sodium lamp operated at 1 kHz with (a) 100% and (b) 25% duty cycle

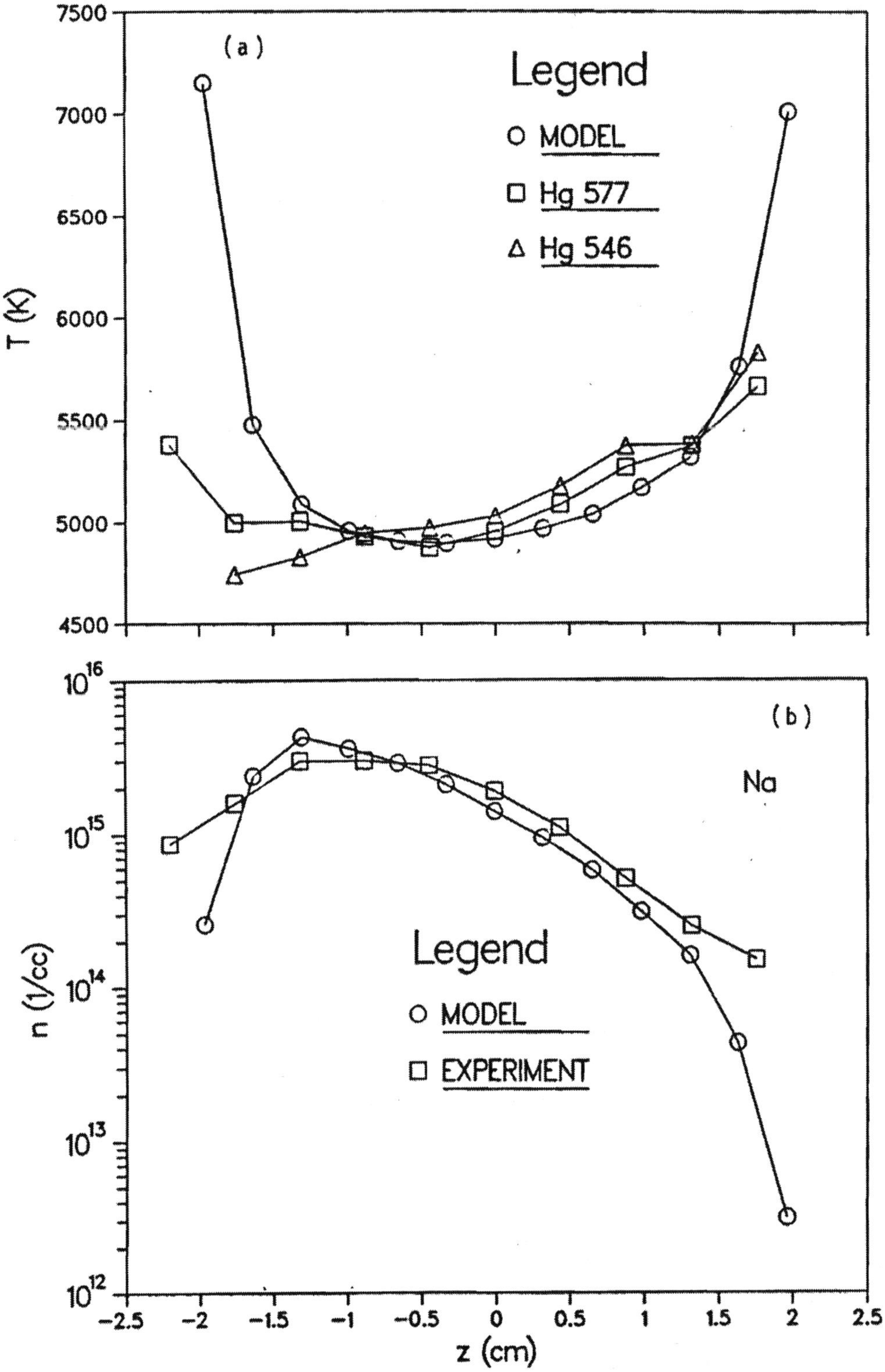

Figure 2: Predicted and observed (a) temperatures and (b) sodium densities along the axis of a vertical metal halide discharge

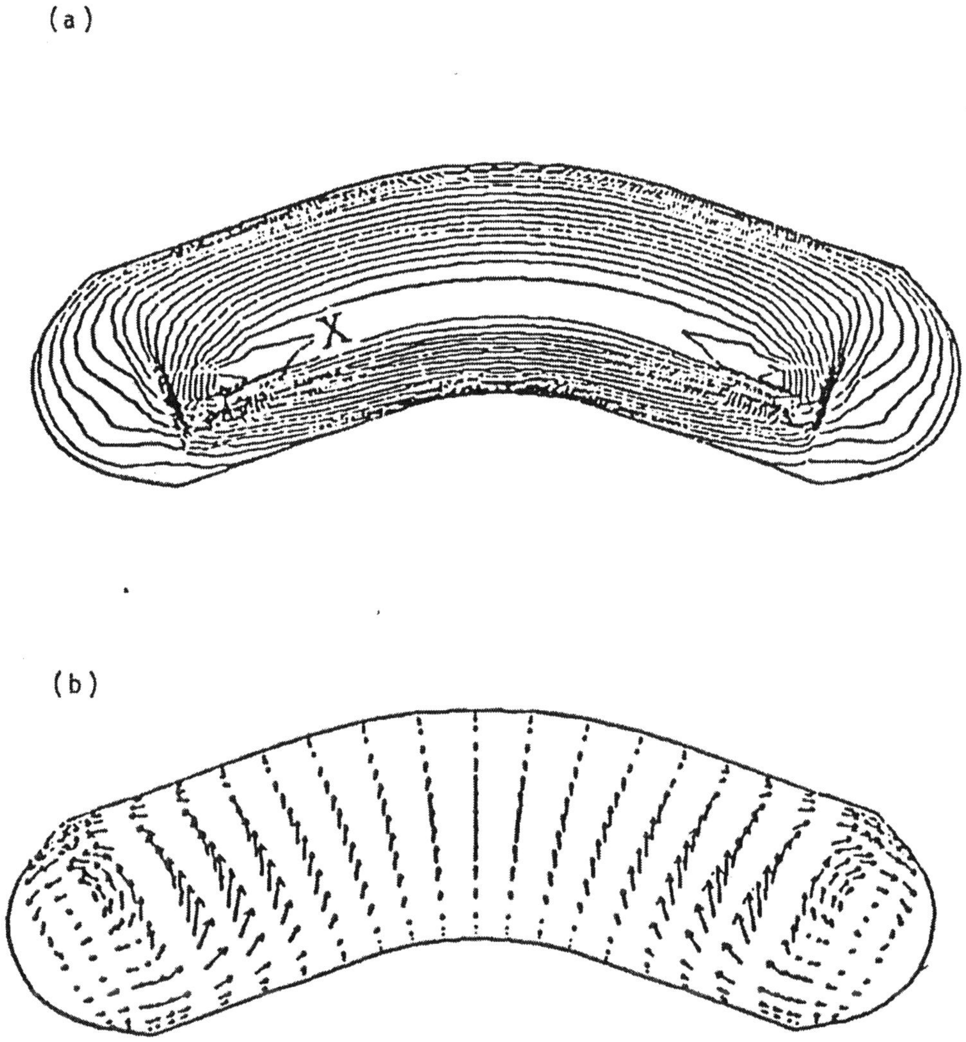

Figure 3: Predicted (a) temperature contours and (b) convection pattern for a horizontal high pressure mercury discharge in a bent arctube.

instance, we have seen the tip of a potential iceberg, but have no comprehensive model or measurement of its size. In the case of cathode phenomena, we have some preliminary models, but few diagnostics of the microscopic phenomena. In the case of life limiting mechanisms, we have some qualitative understanding of mechanisms, but little quantitative understanding of rates.

References

[1] J. deGroot and J. van Vliet, *The High-Pressure Sodium Lamp* (Deventer, Antwerp, 1986).

[2] J.F. Waymouth, *Electric Discharge Lamps* (MIT, Cambridge, 1971).

[3] W. Elenbaas, *The High Pressure Mercury Vapor Discharge* (North-Holland, Amsterdam, 1951).

[4] A.G. Jack and M. Koedam, J. Illum. Eng. Soc., **3**, 323 (1974).

[5] P.L. Denbigh, B.F. Jones and D.A.J. Mottram, J. Phys. D., **16**, 2167 (1983).

[6] D.O. Wharmby, J. Phys. D. **17**, 367 (1984).

[7] R.J. Zollweg, J. Illum. Eng. Soc. **12**, 275 (1983).

[8] J.T. Dakin and R.P. Gilliard, J. Appl. Phys., **60**, 1281 (1986).

[9] J.T. Dakin and R.P. Gilliard, J. Appl. Phys., **62**, 79 (1987).

[10] J.J. Lowke, R.J. Zollweg and R.W. Liebermann, J. Appl. Phys., **46**, 650 (1975).

[11] C.L. Chalek and R.E. Kinsinger, J. Appl. Phys., **52**, 716 (1981).

[12] H.P. Stormberg and R. Schafer, J. Appl. Phys., **54**, 4338 (1983).

[13] J.T. Dakin and T.H. Rautenberg Jr., J. Appl. Phys., **56**, 118 (1984).

[14] C. Kenty, J. Appl. Phys., **9**, 53 (1938).

[15] J.J. Lowke, J. Appl. Phys., **50**, 147 (1979).

[16] R.J. Zollweg, J. Appl. Phys., **49**, 1077 (1978).

[17] W. Shyy and J.T. Dakin, Int. Comm. Heat Mass Transf., **15**, 51 (1988).

[18] E. Fischer, J. Appl. Phys., **47**, 2954 (1976).

[19] H.P. Stormberg, J. Appl. Phys., **52**, 3233 (1981).

[20] R.J. Zollweg, R.W. Liebermann, and D.K. McLain, J. Appl. Phys., **52**, 3293 (1981).

[21] D.E. Work, J. Illum. Eng. Soc., **7**, 118 (1978).

[22] G.R. Allen, R. Lagushenko, J. Maya and W.M. Keefe, J. Illum. Eng. Soc., **16**, 13 (1987).

[23] D.E. Work, Light. Res. Tech., **13**, 143 (1981).

[24] J.J. Lowke, J. Quant. Spectrosc. Radiat. Transfer, **9**, 839 (1969).

[25] B.F. Jones and D.A.J. Mottram, J. Phys. D. **14**, 1183 (1981).

[26] H.E. Lee and L.E. Cram, J. Phys. D. **18**, 1561 (1985).

[27] J.J. de Groot, J. Schlejen, J.P. Woerdman and M.F.M. de Kieviet, Philips J. Res. **42**, 87 (1987).

[28] R.J. Zollweg, R.W. Liebermann and R. Burnham, J. Quant. Spectrosc. Radiat. Transfer **18**, 133 (1977).

[29] H.P. Stormberg, J. Appl. Phys., **51**, 1963 (1980).

[30] A.C. Palacin, Ph.D. Thesis, University Karlsruhe, 1988.

[31] E. Fischer, Philips J. Res., **42**, 58 (1987).

[32] P. Tielemans and F. Oostvogels, Philips J. Res. **38**, 214 (1983).

[33] J.F. Waymouth, J. Light Vis. Env. **6**, 53 (1982).

[34] J.F. Waymouth, J. Illum. Eng. Soc., **16**, 166 (1987).

[35] R. Schafer and H.P. Stormberg, J. Appl. Phys., **57**, 2512 (1985).

[36] D. Karabourniotis, Opt. Commun., **61**, 38 (1987).

[37] D. Karabourniotis, Opt. Commun., **67**, 214 (1988).

[38] J.T. Dakin and W. Shyy, J. Electrochem. Soc., accepted for publication (1989).

MERCURY CONSUMPTION IN FLUORESCENT LAMPS.

P.W.C. Verhees
Philips Lighting B.V.
Eindhoven, The Netherlands

The amount of liquid mercury that is introduced in a (compact) fluorescent lamp (Philips lamps: CFL 5.3 mg ; FL 15 mg) many times exceeds the amount of mercury vapour needed for optimal luminous flux (typical 3 to 200 μg). A reduction of the amount of liquid mercury introduced in a fluorescent lamp becomes increasingly important for the lighting industry.

The main reason is the growing awareness of the authorities, industry and the public that mercury forms an environmental hazard. Since used fluorescent lamps contribute to the mercury pollution in the environment, it is likely that in the near future more rigid regulations concerning the mercury content of fluorescent lamps are established. The best way to decrease the environmental mercury pollution by fluorescent lamps is a substantial reduction of the mercury amount.

A profound insight in mercury loss mechanisms during the operation of fluorescent lamps may enable us to develop lamps in which the introduced amount of mercury is significantly decreased. There has been little experimental work in this area and, consequently, our current knowledge about mercury consumption rates in fluorescent lamps is rather poor. Also the fundamental mechanisms that contribute to the loss of mercury are not well understood.

In the presentation we will give an impression of the present knowledge about mercury consumption using the information available in the current literature.

The interaction of the mercury discharge with glasses has had considerable attention in the literature. It appears that the uptake of mercury by alkali containing glasses (especially soda lime glass) is one of the major causes of mercury loss. Another important mercury consumption mechanism that can be recognized is the reaction of mercury with the various impurities present in fluorescent lamps. Chemical reactions with oxygen- and/or carbon containing species lead to the formation of HgO or carbon-mercury compounds. Also the interaction of mercury with phosphors and with electrode- and emitter materials is shortly discussed.

A variety of possible measures to prevent mercury consumption are suggested. One that will probably be effective is to reduce the role of sodium, which is involved in several mercury consumption mechanisms. This can be achieved by the application of a barrier layer between the glass and the phosphor.

In our laboratory we have carried out experiments using Y_2O_3 barrier layers. It appears that the use of this type of layers is an effective means to reduce mercury consumption. The mercury consumption rate is found to decrease with at least two decades.

COMPUTERIZED TOMOGRAPHY OF HIGH INTENSITY ARCS

T. Fohl and P. A. Vicharelli

GTE Laboratories Incorporated

Waltham, MA 02254

One of the most powerful diagnostic techniques used in the characterization of high-intensity-discharge (HID) arc lamps is that of optically-thin emission spectroscopy. This technique, usually applied to arcs of axisymmetric cross-section, involves the measurement of the radial emission coefficient projected along various lines of sight. Subsequently, the radial distribution that corresponds to these line-integrated projections is obtained through a computation of the inverse Abel transform. Unfortunately, the method is severely restricted in that it is only applicable to discharges that possess radial symmetry, e.g., vertically oriented arc lamps. The purpose of this paper is to describe our experience with a more general approach to emission diagnostics that allows the study of discharges of arbitrary cross section.

The method we use is computerized tomography (CT) of optically-thin radiation emitted from the arc. Tomographic techniques are widely used in medical imaging, in non-destructive industrial inspection, in certain astronomical measurements, and in the study of fusion plasmas.[1] To our knowledge, the present work represents the first application of CT to enclosed arcs such as HID lamps.

The principle of this method is illustrated in Fig. 1. An array of detectors is used to measure the integrated emission intensities originating from lines of sight that lie on some plane of the arc plasma. The same measurement is performed from a series of angles and the results are numerically inverted to yield the internal distribution of emission coefficients for a cross-sectional plasma slice. This distribution is obtained from a solution of the Radon transform, a two-dimensional generalization of the Abel transform.[2] In practice the measurement is made in many planes simultaneously and the CT reconstruction gives the three dimensional distribution of emission coefficients of the arc. From these emission coefficients distributions of temperature and minority species concentrations can be determined.

The essential feature of CT is an axis around which the detector array rotates as it views the arc. The intersecting fields of view of the detector at two orthogonal positions define a volume made up of a sequence of two-dimensional cross-sections, as shown in Fig. 1. The internal distribution of emission coefficients may be entirely arbitrary, the only restriction being that it must be wholly contained within this volume. In our apparatus the detector is mounted on an arm attached to a heavy-duty rotary stage with the axis of rotation passing through the arc tube. The detector assembly is a Vidicon camera with a 75 mm lens. An interference filter with a passband of 1 nm FWHM centered at 577 nm and a neutral density filter are mounted in front of the lens.

The analog signal from the Vidicon is digitized by a frame grabber board in a personal computer. Subsequently, image processing software is used for calibration, and for the actual extraction of data in the form of picture elements (pixels). Typically, rectangular arrays of 100×100 pixels are used with various spacings to define the tomographic volume as the detector array is rotated around the lamp. A typical run involves data acquisition from 36 angles over a 180 degree arc and generates one and a half megabytes of raw data. These data are then transferred to a parallel processing computer, where reconstruction algorithms produce a three-dimensional distribution of 577 nm Hg emission coefficients. A typical reconstruction requires over 60 megabytes of core storage and takes approximately an hour of execution time on a Multiflow Trace 14/200 minisupercomputer. The reason for such an enormous storage requirement is that our reconstruction algorithm has been optimized for stability (it is guaranteed to converge) at the expense of computer memory. Stability is an important consideration because Abel and Radon inversions are very susceptible to noise amplification.

As an example of the results obtained through this technique, Fig. 2 displays the temperature distribution of a 175 W mercury arc operating vertically. The lamp is run on an electronic power supply which maintains a constant magnitude of current but reverses the polarity approximately 540 times/sec. This procedure avoids the necessity of phase-resolving the image acquisition time.

REFERENCES

[1]A. C. Kak and M. Slaney, *Principles of Computerized Tomographic Imaging* (IEEE Press, New York, 1987).

[2]S. R. Deans, *The Radon Transform and Some of its Applications* (Wiley, New York, 1983).

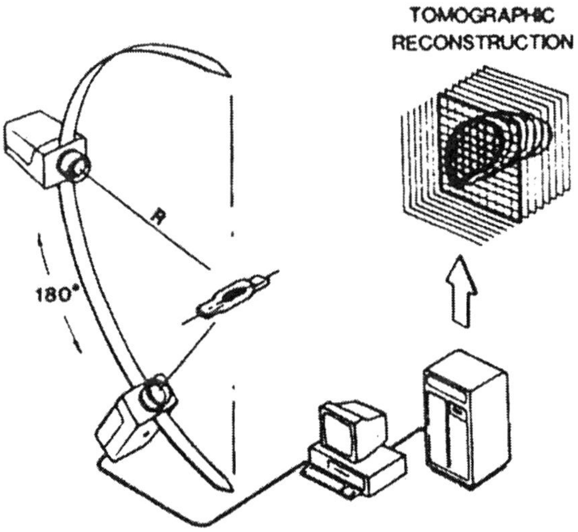

Figure 1.

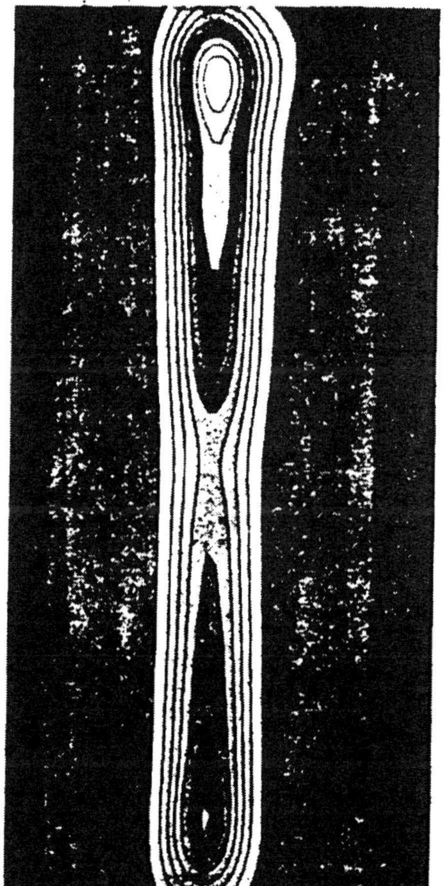

Figure 2.

WHERE WILL THE NEXT GENERATION OF LAMPS COME FROM?

John F. Waymouth, Consultant
16 Bennett Rd
Marblehead, Massachusetts, 01945

In 1979, in commemoration of the 100th anniversary of the commercial introduction of the incandescent electric lamp lighting system, I presented a paper at the Enschede symposium, looking back over the history of the lighting industry to provide insights for the future.

A central theme of that paper was that progress in new light source development has been highly sporadic, concentrated in a limited number of periods of exceptionally high rates of revolutionary new product introduction, separated by periods of consolidation. I called these "Gushers of Progress" (Fig 1). I pointed out that the periods of consolidation appeared to average 15-20 years; within that time frame, a new "gusher of progress" would be expected in the 1980s.

While there has indeed been progress in the 1980's, driven by the energy crisis (which was already full-blown in 1979), I would characterize it more as a "dribble" than a "gusher". Note in Fig 2 that although new products were introduced, they were primarily modifications of existing lamps or technology, new applications of previously well-known developments. This is in sharp contrast to the prior "gusher" periods, marked by the introduction of lamp families embodying totally new principles of operation, with correspondingly dramatic improvements in performance over the prior state of the art. There is nothing in the recent history that remotely approaches the introduction of high-pressure mercury lamps, low-pressure sodium lamps, and hot-cathode fluorescent lamps in the 1930's, or the revolution brought about by the introduction of tungsten-halogen, HPS and metal-halide lamps in the 1955-65 time period.

Similarly, the program of the Karlsruhe symposium three years ago was dominated by reports of research into modelling, diagnostics, and materials. From the topics proposed and the invited speakers for this symposium, I expect that similar research accomplishments will be reported.

This is not in any way to cast aspersions on any of the work. The scope and sophistication of the modelling and diagnostics of metal halide lamps reported is absolutely brilliant. It gives great confidence that we are fast approaching a full understanding of (and therefore the ability to control and manipulate) these complex discharges. The practical performance of all light sources is limited to a very large degree by capabilities of materials; the research into improved materials will surely lead to achieving in the real world the full performance of which these sources are capable.

Nevertheless, these are all "consolidation" research, not revolutionary discovery.

Placing myself for the briefest instant in the shoes of the corporate managers and directors who provide the funds for light source research, I have to make the statement: "If the result of your very expensive research in modelling and diagnostics is merely to confirm that these products are pretty well optimized and that only marginal improvement in performance is possible, then I am not interested in funding it any longer. The world does not need another formula of metal halide lamps of 10% performance improvement, unless they operate in existing sockets on existing control gear, can be manufactured on existing equipment using existing processes, and can be introduced at only modest cost."

The truth of the matter is that however much lip service captains of industry give to R&D, they abhor change. Change embodies risk, and requires decision. New products of only modest improvements in performance, but requiring major investment in manufacturing equipment and new-product introduction costs are unwanted stepchildren in the corporate nursery. They are routinely strangled somewhere between the laboratory and the factory. Any corporate mogul with his eye on this year's bottom line will infinitely prefer to invest in making his existing product better, faster, and cheaper, so he can discount it further while making more money.

The unique characteristic of the major products introduced in the "gusher" phases of prior history is that they literally forced their way into the marketplace, because they had performance more than double the prior art. Not a single lamp maker dared commit infanticide on any of them, for fear that others would successfully introduce the product, obsolete his existing product portfolio, and siphon off his market share. In any number of cases, the decision to proceed was a classic case of "bet the company". Introduction costs were so high that it was a decade before a profitable return could be expected, but failure to get on the bandwagon in time meant market surrender.

Because none of the developments since metal halide, with the possible exception of selective-filter incandescent/T-H, have had a benefit/cost performance advantage of a factor of two over existing products, none have had the revolutionary impact of fluorescent, T-H, HPS and MH lamps.

Hence my question, "Where will the next generation of lamps come from?" really means "When will the inventors, engineers, and scientists force the lamp companies from their comfortable oversight of a well-understood product and market spectrum into the management of near-catastrophic change?"

The answer is, of course, not until we begin to look farther afield than the present systems of converting other forms of energy into light. Nothing reported at this meeting has the capability for a factor of two or more improvement.

In the presentation, I will suggest some areas of investigation that might have such a potential.

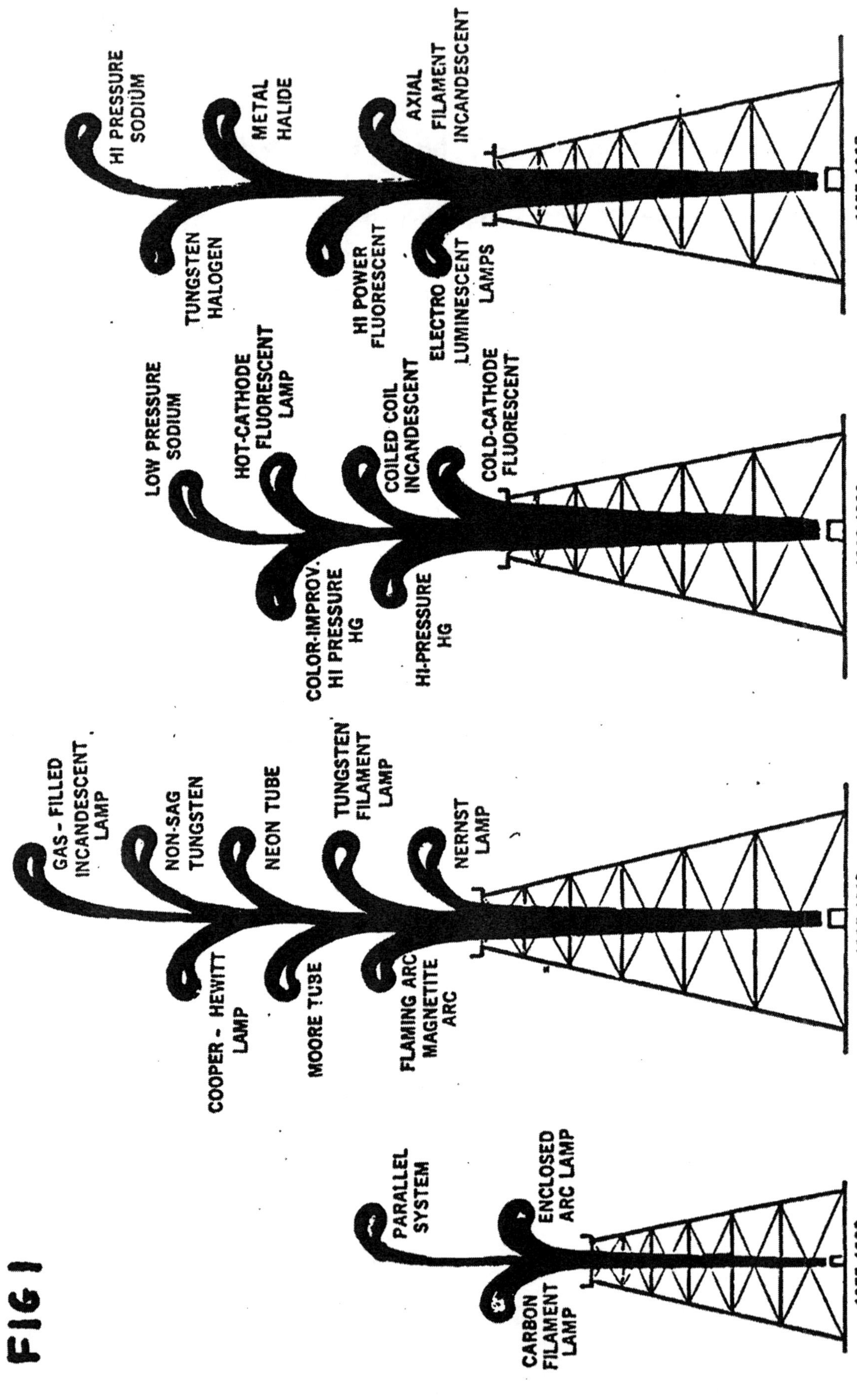
CARBON FILAMENT LAMP
PARALLEL SYSTEM
ENCLOSED ARC LAMP
1875-1886
COOPER - HEWITT LAMP
MOORE TUBE
FLAMING ARC MAGNETITE ARC
GAS - FILLED INCANDESCENT LAMP
NON-SAG TUNGSTEN
NEON TUBE
TUNGSTEN FILAMENT LAMP
NERNST LAMP
1897-1913
COLOR-IMPROV. HI PRESSURE HG
HI-PRESSURE HG
LOW PRESSURE SODIUM
HOT-CATHODE FLUORESCENT LAMP
COILED COIL INCANDESCENT
COLD-CATHODE FLUORESCENT
1930-1938
TUNGSTEN HALOGEN
HI POWER FLUORESCENT
ELECTRO LUMINESCENT LAMPS
HI PRESSURE SODIUM
METAL HALIDE
AXIAL FILAMENT INCANDESCENT
1955-1965

FIG I

FIG 2

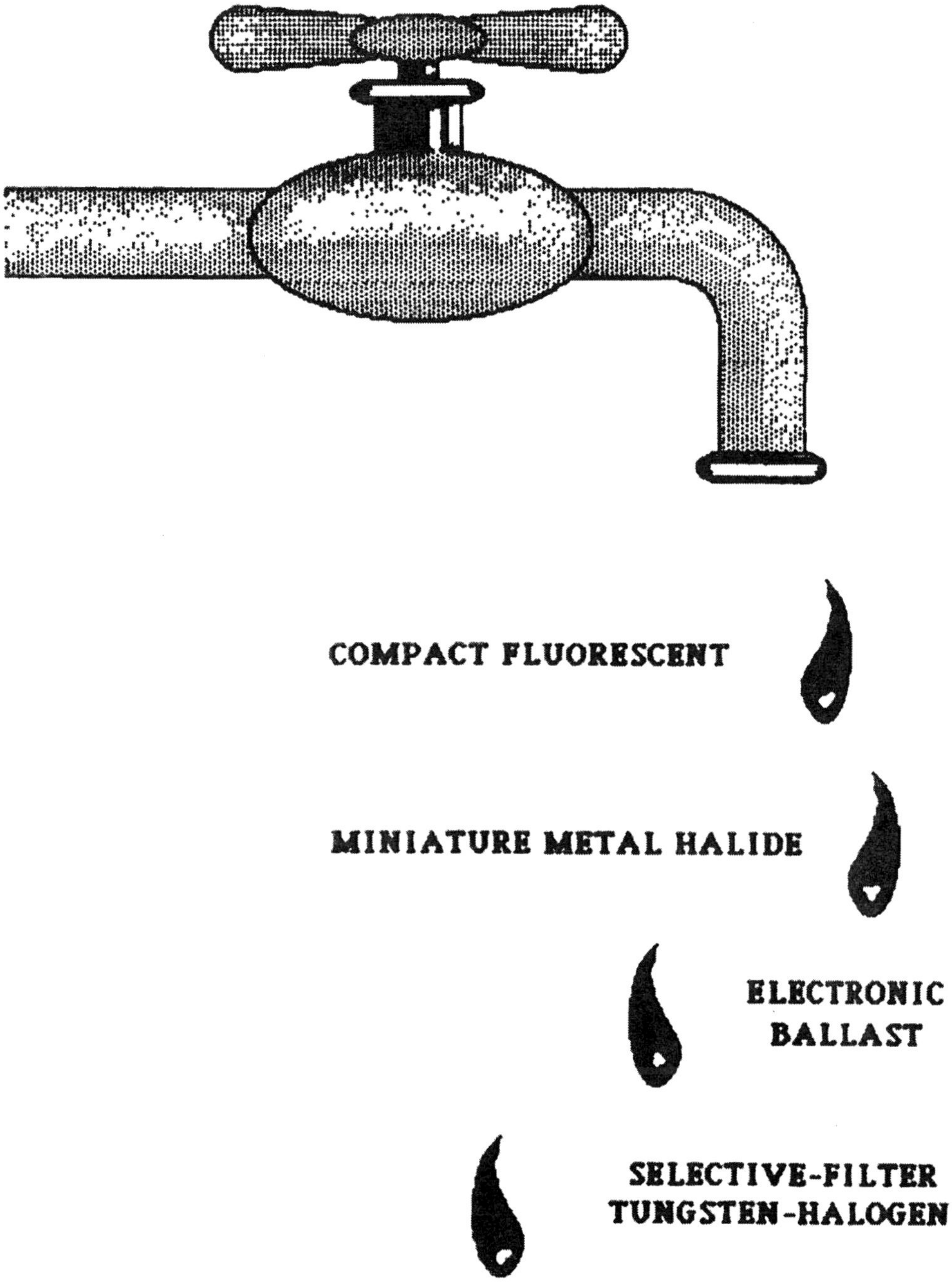

DRIBBLES OF PROGRESS
1975-1989

CHARACTERISTICS OF POWER SUPPLIES FOR STANDARD LAMPS

Author: E.M. Wareham C.Eng., M.I.E.E

Address of Laboratory: Vinculum, 40a Kneesworth Street, Royston, Herts SG8 5AQ

The development of low voltage high current lamps such as the NPL/GEC Clark-Berry heavy duty standard improved the stability of standard sources to the point where minute changes in the circuit contact resistances resulted in unacceptable deterioration of the lamp performance when powered from a constant voltage source. Constant current supplies stable to better than 0.01% were developed to deal with this problem, the lamps being calibrated in terms of the filament current.

When both the current and voltage can be measured with precision, the output of the power unit can be made to control a mathematical relationship between these two parameters. For example, if their product is held constant it is the power into the source which is defined independent of changes of resistance. Similarly, with materials such as Tungsten which have a very large temperature coefficient of resistance, the ratio of voltage to current defines the resistance of a lamp filament which is directly proportional to its mean temperature.

This paper considers these and other alternatives and their relevance to light sources of different kinds powered from stabilised AC and DC supplies.

Author Index

Adam J 56:P
Albach M 37:P
Alkaa A 16:P
Altena F W 18:L
Anzai Y 3:P
Ascher R 31:P
Aubes M 70:P
Awadallah A S 67:L
Azinovic D 81:P
Bencze L 68:P
Beneking C 38:P
Benetruy P 15:P
Bergervoet J R 52:L
Bergman R S 65:L
Berman S 43:P
Bhattacharya A K 67:L 69:P 72:P
Bigio L 55:P
Biro E 56:P
Bolte E 53:I
Boomkens A 18:L
Brumleve T R 30:P
Cao S C 64:P
Chang P Y 86:P
Chittka U 73:P
Cifuentes L 87:P 88:P 89:P 96:P
Couris S 70:P 90:P
Curtis E H 71:P
Dadonov V F 4:P
Dakin J T 86:P 104:I
Damelincourt J J 15:P 16:P 90:P 91:P
Datta R K 21:P
Davenport J M 85:L
Devonshire R 23:P 25:P 57:P
Dietz K 47:P 92:P 93:P
Dittmer G 22:P
Dohn A 47:P
Drakakis E 70:P 91:P
Duffy M E 86:P
Duncan M 23:P
Edwards A L G 45:P
Edwards H G M 25:P
Eliasson B 39:P 51:L
Fawcett V 25:P
Fedorenko A S 4:P
Fields M 24:P 25:P
Fijan D 33:P
Fischer E 20:I 26:P
Foerster D 7:P

Fohl T 103:L 106:L
Gambar K 77:P
Gavrilkina G N 99:P
Gellert B 39:P 51:L
Godyak V A 40:P
Goodman T M 109:P
Greb U 9:P
Green D G 57:P
Greve P 41:P
Groves R D 72:P
Guenther K 42:P 58:P
Gundel H 13:P
Hansler R L 85:L
Heuvelmans J J 79:P
Hilbig R 73:P
H[illegible]pert K 27:P
Hollister D 43:P
Hollo S 74:P
Hope L L 5:P
Husgen T 49:P
Imai T 50:P
Ingold J H 69:P
Ivantzev A S 59:P
Jia G X 48:P
Jonas S 28:P
Kaiser R 26:P
Kanabus T 60:P
Kaposi O 68:P
Karabourniotis D 75:P 102:P
Kastelein L 17:L
Keeffe W M 29:P 94:P
Keijser R A J 84:L
Kindel E 6:P
Kloss H G 42:P 58:P
Koedam M 1:I
Kogelschatz U 39:P 51:L
Kokinov A M 76:P
Konjashkina V I 59:P
Korochkov J M 76:P
Kreher J 7:P
Kronert U 47:P
Kwatera A 28:P
Lange H 6:P
Lantos F 8:P
Lehmann T 42:P
Lesiuk B 28:P
Liebermann R W 94:P 95:P
Ligthart F A S 17:L
Lister G G 9:P
Lohmann G 44:P
Lojka O 56:P

Luedge W 58:P
March P 11:P 12:P
Markus F 77:P
Martjinov V I 4:P
Mascarenhas E J P 45:P
Mayer G 47:P
Meads J 23:P
Meier W 7:P
Meyer V D 29:P
Miller M 27:P
Moore J R 109:P
Mordjuk A V 62:P
Mottram D A J 87:P 89:P 96:P
Mouse M 110:I
Mucklejohn S A 23:P 24:P 30:P 57:P 78:P 90:P
Muller H 46:P
Murayama S 19:I
Nagel F 83:I
Narita M 97:P
Neiger M 43:P 44:P 46:P 47:P 98:P
Niemann U 22:P 26:P
Nyiri B 10:P 74:P 77:P
O'Brien N W 23:P 24:P 30:P 57:P
Obmajikina M I 59:P
Osawa T 3:P
Palacin A C 98:P
Palladas A 75:P
Pichler G 33:P 81:P
Popovic A 68:P
Popp H 31:P 44:P 47:P
Por A 15:P
Prikupets L B 99:P
Prjitkov A A 4:P
Ptak W 28:P
Pustoshkin A A 80:P
Qin K J 48:P
Radtke R 42:P
Ranish J M 63:P
Reichert T 47:P
Reznik S 56:P
Roaux E 91:P
Rogajilin M I 62:P
Rothwell Jr H L 94:P
Rowley A T 78:P
Rukavina J 33:P 81:P
Sai-Baba M 27:P
Saikatsu T 3:P
Sakurai T 3:P
Salon J 91:P
Sarytchev G A 99:P
Schenkelaars H J W 79:P

Schlejen J 79:P
Schnedler E 20:I
Schonlein A 92:P 93:P
Schorpp V 46:P
Schulman M B 82:L
Schwarz W 11:P 12:P 93:P
Serick F 42:P
Shinn D B 35:I
Shyy W 86:P
Smorodinova L A 99:P
Snellgrove R 34:L
Solleveld G D A 32:P
Sonnenberg R 13:P
Soules T F 69:P
Springer R H 85:L
Stanioch W 28:P 60:P
Stockwald K 46:P
Szymanski A 61:P
Ter Heyden H 17:L
Tsakonas A 75:P
Van der Kooi A G 71:P 79:P
Van Pijkeren D 100:P
Van Rijswick M 18:L
Van Vliet J A J M 101:P
Vasquez S 87:P
Vdovin N S 62:P
Vdovin V G 80:P
Vdovina V A 80:P
Verbeek W 101:P
Verhees P W C 105:L
Veza D 33:P 81:P
Vicharelli P A 102:P 103:L 106:L
Walasek E 28:P
Wang L C 48:P
Wani K 2:L
Wareham E M 108:P
Waymouth J F 107:I
Wegener A 49:P
Westerhof M 18:L
Wharmby D O 36:I
Woodward D R 82:L
Woodward J R 54:L
Wyner E 34:L
Yang L X 48:P
Yoshikawa N 50:P
Yousfi M 16:P
Yu T H 63:P
Yuge Y 66:I
Zhongliang M 14:P
Zhou W Z 64:P
Zissis G 15:P 16:P 90:P